Sex in Space

by

Laura S. Woodmansee

Also by Laura S. Woodmansee:

Women Astronauts
Women of Space: Cool Careers on the Final Frontier

Published by CG Publishing Inc, Box 62034, Burlington, Ontario, Canada, L7R 4K2
Printed and bound in Canada
©2006 CG Publishing Inc/ Laura Woodmansee
Sex in Space/Laura Woodmansee
ISBN 1-894959-44-2
ISBN13- 978-1894959-44-5
Cover by Geoff Godwin

Sex in Space

By
Laura S. Woodmansee

Dedication:

To the Next Generation, especially Nathan,
may you boldly go where no one has gone before.

Table of Contents

Foreword / Foreplay
by Rick Tumlinson,
President of the Space Frontier Foundation

We are all in the gutter, but some of us are looking at the stars. ~ *Oscar Wilde*

From Helen of Troy to Caesar and Cleopatra, to Bill and Monica, wars have been fought, empires broken and regimes changed because of love and sex (yes Bill, it was sex). In fact, human sexuality and romance has been at the root of many historical changes. It is a real and yet often denied undercurrent in all aspects of life. It is this very real and "human" side of ourselves that adds the spark to the engine of history. It is also the engine of a large part of our economy.

In financial terms, sexuality and romance drive huge sectors of our society, producing trillions of dollars of activity. Think about it, be it overt or covert, sexuality and the partnering dance is big business. Whether it be mascara or music, sappy Valentine's cards or sex sites on the internet, the drive to be sexually attractive, sexually fulfilled, to love and make love, to find the right mate to better copulate is all around us. And let's be real here, the darker side of sex is also one of the biggest industries on Earth. Home video exists because people could suddenly watch pornography without going to a "perv" filled theater. Cable and satellite television were driven in large part by their ability to broadcast sexual content free of government interference. The Internet itself, the most revolutionary technology since the telephone, although begun as a tool for government researchers, took off when someone came up with the idea that people could have sexual interactions without leaving the comfort of their own homes.

The influence of sex on our society goes far beyond products and programming. Entire cultures and national economies are based on these pursuits. Lose the sexual imperative and every nightclub on Earth is closed the same day, along with half the stores at the mall, take the sexual dance out of Brazil's carnival and you get some priests waving incense around and shuffling down the streets of Buenos Aires, lose the lasciviousness of Las Vegas and you end up with a few robotic old ladies staring into slot machines, take the sexual dance off the beaches, and half the resorts in the world would close tomorrow. Yet, there are those of narrow mind and mental immaturity, whose personal issues have been translated into religious zeal, who would end this discussion in a flash, as well as any other that even admits such things are possible. Throughout our history, as important as it is to the human condition in all its aspects, the repression of the sexual dance has and is the driver of darkness and fear.

And so, of course, as we enter the new frontier of space, we take the conflict and hypocrisy of the sexual debate with us. Our space program, born of our military, an even stiffer subset of our already semi-puritanical male dominated American society, of course reflects the Puritanical fanaticism of some of our founders, wherein any mention of human sexuality is verboten! The idea that anyone working for the space program is a sexual being, or that somehow someone sometime somewhere might have crossed the line and explored one of the root drives of being human is a non-starter in official space circles. Yet the rumors fly - so to speak. Call it one of the first "spacer legends" if you will. After all, surely someone has "done it" in space. Was it two very bored, horny and sexually experimental cosmonauts finally breaking down after six months locked in a tin can together? Was it a married crew on the space shuttle sneaking about for a little private debriefing in the main deck after lights out? We may never know for sure when this first happened, or if it has but if it hasn't, it will very soon. (I am sure Laura will be filling you in on this in the following pages.)

Sex in space is coming, so to speak. As we move into the era of NewSpace, when private operators are developing commercial spacecraft and even space hotels, one of the activities sure to be on their guest's agenda will be - you got it - joining the 100 mile high club! It's going to happen, it's going to happen a lot, and in fact, just like Vegas or the Bahamas, it will be a must-do for any couple venturing upwards. I will wager that the first "honeymoon" in space is only a couple of years away, if it is only for a few minutes on a Virgin Galactic spaceliner or some similar craft (plenty of time for most males according to Cosmo Magazine), followed shortly thereafter by a real honeymoon in an orbiting hotel. And soon enough the Moon in honeymoon will be literal, as lovers consummate their affections as their ship flies above the mountains of Luna. The imagination soars at the possibilities, and the questions flow. Mechanics and mentality, leverage and love. And then what? As space flight becomes routine, human beings will not only experience the joys of making love in 3 dimensions, but will they, along the way create entire new dimensions for the expression of passion?

Of course, where there is sex, there is also procreation, and at some point, the first baby conceived in space will be born. Or will it? We don't know if a human baby can be conceived in space, although some experiments have been done. What happens to sperm and egg in a micro-G environment? What happens during cell division? Does the growing fetus develop bone structures or a strong enough vascular system to deal with gravity? Can a child be conceived, carried to term and born in space ever return to the Earth? What of one born in the 1/3 g of Mars, or the 1/6th of the Moon? How much gravity is needed for a baby to develop in such a way it can return home? Or are we on the verge of creating new branches of humanity specially conformed for life in space, on the Moon and Mars, who will not only never return to the Earth, but may not wish to due to its crushing gravity? The research has been done here and there, but what is the over all picture? What should we be concerned about and what is purely speculation and fear mongering?

Thus this volume. The author has done a skillful job of lifting the covers of this mystery for us, which has long needed to be done. And she does it with taste - although in language that is frank and very direct. And this is important, as the subject deserves no veil of fear, nor should such discussions be censored. There needs to be a clear separation for example, between the act of sex in space and procreation in the micro gravity environment, as the consequence of one is a great memory, and the other, a living being.

Laura has begun a discussion here that will be with us a long time. And it is a discussion we need to have. It is important that the author is a woman, as she can speak of the subject in a tone no man could muster. Given the way men have handled the subject of sex throughout history, I suggest it is about time someone else lead at least part of the discussion, and this is the right way to begin. Be it the culturally enforced wearing of buhrka by the Taliban, or the ongoing battles over the teaching of sexuality in western cultures, there seems to always be someone who feels this most natural of all human activities is somehow wrong. Interestingly, they are usually males. After all, most of the tenets of fundamentalism, or today's fanatic religions promoting sexual repression and social stigma were and are inventions of the male gender - not the female. It is men who drive the social systems that force the covering of faces and bodies lest lustful thoughts occur (in men!). Even more interestingly, as scandal after scandal reveals, they are often the same males who can be found secretly perusing the Internet at night looking for sex.

The darkness of ignorance so often surrounding the subject of sex must not be allowed to follow us into space. I can easily see that from a feminine perspective the whole field must seem funny. A bunch of geeks building giant cylinders which they use to "boldly thrust into space" ...nuff said.... With books like this, perhaps the newest wave of space women can begin to soften the edges of this male dominated field. A softer, more embracing view of space tilts us away from the macho "right stuff" imagery of hard nosed test pilots stepping from fighter jets to rocketships and attacking the frontier, to the idea of a human community expanding outwards as a natural extension of what it means to be alive. Women have been a part of the opening of space from the start. Unfortunately, they have been treated as second-class citizens, worse, they have been treated as men, not the separate and wonderful beings they are. Yet women bring a new dimension to this amazing frontier, if they are allowed to be themselves. We now have shuttle pilots and astronauts of all sorts who are women, working side by side (if not hand in hand) with men. It is time to let them be women, and by extension, to let the men relax a bit too. The male hypocrisy that has characterized our cultural sexual dysfunction should be left on the savage savanahs of this world, with the chest pounding apes who promulgated it. The yang of male dominance in space needs to make way for the yin, the soft side.

Let us enter this new frontier with our eyes - and our minds - open. Let us welcome the dance of sexuality and love, not deny it. Let this new attitude feed upwards into the larger picture, as we move our mindset regarding the frontier from conquest to dance... Perhaps, if given the chance, the next generation of visitors and explorers in space can help create the new language of space - love. But most importantly, let us have fun! It is time to get this party started!

Introduction: What kind of a book is *Sex in Space*?

"There is nothing more powerful than an idea whose time has come." - Victor Hugo

Throughout *Sex in Space*, I've tried to be open and honestly discuss space sex and the issues related to it, such as who has done it, sexual positions in weightlessness, reproduction, what space is like, space tourism, and many other topics. Obviously, sex is a subject that causes a lot of embarrassment but as my father says, "It's the most natural thing in the world." He's right!

This is why you don't need to be embarrassed or ashamed to pick up this book and read it! I've created an "embarrassment-free zone" within the covers of this book. *Sex in Space is* intended for mature people, but is not pornographic. On the other hand, if I make you blush or giggle at times, that's ok; it's part of what makes us human.

My husband jokingly calls this a pleasure book. What he really means is while *Sex in Space* deals with a serious topic, it reads as if I were talking with my best friend. I write in the first person, use sexual innuendo, include rumors, and let my imagination run wild, but I clearly identify what's what.

This is not a textbook, but amid the fun there are a lot of hard facts to better explain what space is like and how it relates to sex. I had fun learning about the subject, talking with experts, and writing about it for you. My goal is for you to have fun reading it too.

When I told a good friend of mine that I was writing a book about sex in space, he said, "Laura, you are the last person that I'd expect to write a book about sex in space, but the best person to write it." This was a great compliment to me. After reading the book, I hope you will agree with him.

Space is inherently sexy and romantic. There's definitely an exhilarating feeling when looking at the beautiful stars of our night sky. Just ask anyone who has gone camping far from the city lights on a dark night, especially with someone they love. Perhaps part of the awe and wonder is the vastness of the unknown and our desire to explore and see it all. Sex in space pushes our primal buttons for sex and exploration of the unknown.

I decided to write *the* book on sex in space because I realized how little attention is paid to this important topic. There are a number of articles, and reports about it, but no one has put the entire story together until now. Everyone who is interested in space has wondered, at least once, what it may be like and if anyone has 'done it.' Since no one has yet done a book on sex in space, I guess that it's up to me to get the ball rolling; or rather to have the balls to get it rolling. Obviously, some basic questions arise when one thinks about the possibilities of sex in space. First off, has anyone had sex in space? There is a lot of fact and fiction surrounding this topic and I discuss it all in the first chapter.

In chapter two, I discuss how people might make love in space. To understand how to have sex in space, we need to better understand what space is like. Once that's clear then we can address what the accommodations might be like, what to wear, some positions that should work and some that may not, and finally what aids, or sex toys, might be especially useful in space.

The third chapter is about reproduction in space. This is a bit more serious than the previous chapters. We'll look at how space affects both male and female reproductive biology, then whether reproduction is possible in space, some difficulties, and what might be the results of conception, pregnancy, and birth in space.

In chapter four, we'll discuss psychological issues that affect people in space, especially on long-term missions, and the implications for sex, both in terms of gender relations and sexual liaisons in space.

We'll take a peek into the future in the last chapter. We'll find out what's happening today in space tourism and what the future may hold for romantic space development. We'll also find out who is working on cheaper launch vehicles, and what a future space hotel might offer its guests. Finally, we'll discuss sex and science fiction, and our distant future in space.

We are entering an era of space tourism and longer-lasting spaceflights. Sex isn't anything to be ashamed of, although it makes most of us giggle. It is part of human nature, and the fact is that sex is the core of what makes us human. We may not like to admit it, but the sex drive rules all. Wherever humans go, sex and reproduction follow, so we'd better be ready to deal with the consequences of the nature of humanity.

Most people who read this book will do so just for fun, or because it's interesting, and that's great. But, I hope this book will also be a useful guide for couples interested in space tourism, what to bring on a space tourism adventure, future business opportunities for space entrepreneurs, inspiration to science fiction writers, and help developers plan for space settlement in the distant future.

Recently, I had the good fortune to attend a space entrepreneurship conference called "Space Billionaires," hosted by my Alma Mater, the University of Southern California (USC). I was amazed by how many times the speakers and attendees talked about the idea of sex in space, or said that they wanted to "make space sexy," to the general public. Sex in space is the "Killer App" of the space tourism business.

In the not-so-distant future, human beings *will* live and work outside the Earth's atmosphere. The time has come for potential space tourists and the builders of future space communities to start thinking about *Sex in Space*!

I hope that you enjoy *Sex in Space*.

Chapter 1: Has Anyone "Done It" in Space?

The 100 Mile High Club

You've probably just picked up this book and are thinking; 'Cut to the chase, has there been sex in space or not? Don't keep me in suspense. Tell me!'

Well, the answer is intriguing, complicated, and will take a bit of explaining. You see, there is no definite *documentation* of sex between humans in space. On the other hand, there are documented cases of sex between animals in space, and between a man and woman in zero gravity. Keep reading and I will reveal all of the intimate details to you.

There have also been a number of rumors and speculation *about* people having sex in space, but the space agencies are not admitting it, or talking about it *directly*. So we need to look at the evidence we do have, and that's what this first chapter of *Sex in Space* is all about. Has anyone 'done it' in space? The short answer is ... well, maybe.

From Russia with Love

The first men and women to fly into space did so during the height of the Cold War between the United States (U.S.) and the former Soviet Union (USSR). At the same time, a sexual revolution was occurring in the United States.

The USSR was the first nation to put "mixed" crews of men and women into Earth orbit. So it seems appropriate that stories of sex in space began with rumors of love among the cosmonauts.

The first woman to fly into space was USSR Cosmonaut Valentina Tereshkova. Her 1963 solo flight lasted three days. There are conflicting rumors that the staff of mostly men could not stand the idea of a woman in space. For unclear reasons, the Soviet Union did not send another woman into space until 1982, just 10 months before Astronaut Sally Ride became the first American woman in space on June 18, 1983 onboard the *Space Shuttle Challenger.* Keep in mind that during the Cold War, the U.S. and USSR were constantly trying to outdo each other.

USSR Cosmonaut and Engineer Svetlana Savitskaya became the second woman to fly into space in August of 1982. Savitskaya flew with

Valentina Tereshkova

fellow Cosmonauts Leonid Popov and Alexander Serbrov. Their *Soyuz T-7* space-craft docked with the Soviet space station *Salyut 7*. Two long-duration cosmonauts, both men, were already onboard the station.

So, the real question here is, did the Soviet Union want to use this first oppor-tunity to experiment with the sexual union of a man and woman in space? The rumors have been flying ever since the trio's 1982 spaceflight. But according to my research, these are simply rumors based on speculation from people who heard that one woman spent 8 days with some men in zero-g onboard a space station.

If you were told about a woman and two men who went on a weeklong camp-ing trip together out in the wilderness, you might think nothing of it. But then again, you may wonder if something romantic or sexual happened. Even if you know nothing about these people, there's bound to be some speculation about pos-sible intimate relations between them. It's not fair, but it is human nature to won-der about these things. That's why you're reading this book, right?

Let's look at the evidence. The *Soyuz* spacecraft is a cramped capsule about six feet in diameter with three bucket seats, instruments everywhere, and not much room to move around. With three people crammed in there, most people would agree that it's no place for a sexual liaison. Although, those who've managed to make love in the back seat of a compact car may disagree. But, there probably was-n't a third person looking on, which is what it would be like in a *Soyuz* capsule. For most people, this would kill the mood, but not for everyone.

A *Soyuz* Spacecraft as seen from Apollo prior to rendezvous

Consider the fact that about half of all space travelers expe-rience Space Adaptation Sickness (SAS) for the first day or two of spaceflight. Nausea, vomiting, sinus congestion, and headaches don't usually induce a romantic mood. This means that about half of any potential orbital lovebirds would almost surely have to wait a day or two, about the time it would take to travel and dock with a station or hotel, to be in any kind of shape to fool around. We'll talk more about this in the Chapter 2, *How to Make Love in Space*.

The USSR's *Salyut 7* space

station was a 44-foot long cylinder with a diameter of about 12-feet, but those were the outside dimensions. With equipment and supplies, there wasn't much free space on the inside. Several people could live and work on the station and, although not exactly spacious, it was not nearly as close quarters as a *Soyuz* capsule. Two *Soyuz* spacecraft docked to the station effectively made a couple of small private rooms. Now there was some space and privacy available for a little hanky panky.

When the three cosmonauts from *Soyuz T-7* arrived at the *Salyut 7* space station, they were greeted by Cosmonauts Anatoli Berezovoy and Valentin Lebedev, the two men who made up the EO-1 crew. That made five people who had to live and work together in a small, confined space. The crew of *Soyuz T-7* brought science instruments, supplies, and mail from home.

Soviet Space station Salyut 7

Once together, the matter of personalities comes into the equation. In recently released letters from Cosmonaut Anatoli Berezovoy to his wife, translated by *Quest Magazine: the History of Spaceflight*, he says, "Serebrov and Savitskaya are fighting like a cat and a dog. Sasha (Serebrov) has turned out to be one of those people who likes to whisper gossip about their friends … Savitskaya is not exactly going out of her way to do anything over and above her program description." This is not exactly the description of a close crew with members who are likely to become romantically, sexually, or otherwise intimate.

After the flight Savitskaya said, "They greeted me at the hatch with an apron." She then explained how she set them straight and established a professional working relationship with her fellow cosmonauts.

Even if the crew did get along well, Savitskaya does not appear to be the type of person who would be willing to have casual space sex. Savitskaya's father Yeveniy Savitsky, was the Deputy Commander of the Soviet Air Defenses, a World War II air ace, and twice Hero of the Soviet Union. Savitskaya was well educated, and a skilled aerobatic pilot who soloed at age 15. She probably owes part of her success to her father's influence, but the majority to her own natural flying ability and perseverance. An *Izvestiya* article from 10 August 1982 reported that "pressing" is her favorite word because she has never been passive, and pressed for every break she got.

Cosmonaut Svetlana Savistkaya, first woman to fly in space with male crew members

This certainly doesn't sound like a stereotypical passive sex kitten. Svetlana Savitskaya had influence, power, and was a strong-willed professional. If a romantic liaison occurred, then it would most likely have been at her instigation. And given her father's position, it is doubtful that any one of the men would have been willing to take that risk, even if they were willing to be unfaithful to their wives. Besides, Savitskaya was, and still is, married to engineer and pilot Viktor Khatkovsky of the Ilyushin Aircraft Design Bureau.

Savitskaya was given the *Soyuz T-7* capsule to sleep in by herself, but according to Berezovoy, she chose to sleep close to the others so as not to be treated differently. Could Savitskaya have been intimate with one or more of these men in the privacy of the *Soyuz* capsule while the rest were on the *Salyut 7*? Yes, it's possible, but highly doubtful, and it certainly would not have remained secret for all these years. Others onboard would have known about it, and someone would have spoken about it, especially if Serebrov was a gossip as Berezovoy claims. All four of the men were married with children and all were Soviet stalwarts cleared by the KGB, which rejected over half the cosmonaut recruits. This doesn't make them uninterested in sex, just less likely to do something to damage their careers and/or marriages.

In July 1984, Savitskaya returned to space on *Soyuz* flight T-12 with Commander Vladimir Dzhanibekov and Igor Volk, on another mission to the *Salyut 7* space station. At *Salyut 7* they met the three-man crew of EO-3 Leonid Kizim, Vladimir Solovyov and Dr. Oleg Atkov. The station was even more crowded than on the previous mission. While on *Salyut 7*, Savitskaya and Dzhanibekov conducted a three-and-a-half hour space walk, or Extra Vehicular Activity (EVA). It was the first time a woman did an EVA. During the space walk, they tested tools for space by cutting, welding, soldering, and coating metal samples outside of the spacecraft. The six cosmonauts aboard *Salyut 7* also conducted resonance tests and collected station air samples.

There is no documentation of any other biological experiments with human sexual reproduction as some tabloids have speculated. For the same reasons given above about Savitskaya's previous flights, it seems ridiculous to assume that she was involved in any outer space affairs on this flight either.

It's interesting to note that Savitskaya was scheduled to command an all-woman crew to *Salyut 7* on the occasion of International Woman's Day. Unfortunately, problems with the station and a lack of available *Soyuz T* spacecraft made the flight impossible. If the flight had taken place, no doubt there would also be rumors about the sexual antics of an all-female crew.

The U.S. National Aeronautics and Space Administration (NASA) toyed with the idea of an all female crew for the Space Shuttle in 1999. Astronaut Rhea Seddon, one of NASA's first batch of women astronauts, led the study. The space agency scuttled the idea after studies suggested the estrogen-laden "all maiden" voyage would yield no scientific value. Perhaps ridicule by the media had something to do with NASA's decision.

What about other USSR spaceflights? Well, the Soviets gave rides to two other women, but they were not cosmonauts, and they were not citizens of the USSR either.

The first was British Chemist Helen Sharman. In 1991, Sharman won a corporate-sponsored trip via *Soyuz MR12* to the *Mir* space station. She was onboard for a week, where she did some experiments with crystal growth in space. A few tabloids at the time made some pretty outrageous suggestions about her flight. Sharman herself told reporters that she had a "fantastic experience" with the cosmonauts onboard *Mir*. An alleged videotape from the mission shows her wearing a pink evening gown. This seems to be joke on her part because an evening gown is obviously quite impractical for a space mission, even if one was interested in space sex. If authentic, Sharman was probably using her comments and dress to make fun of the tabloids.

In Chapter 2, *How to Make Love in Space*, we will take a much closer look at what to wear for that special rendezvous in space.

In an interview with an anonymous Russian official from *Mir*'s Psychological Support Service, Peter Pesavento in his article "The Psychological and Social Effects of Isolation on Earth and in Space" wrote, "Dr. Sharman was very cool and distant personality-wise, and she wanted to be regarded as a scientific specialist first of all – which did not fit the expectations of

British Space Tourist
Helen Sharman

the male cosmonauts she interacted with." This clash of cultures and emphasis on Sharman's professionalism leads me to believe that her flight was all business and any rumors to the contrary were strictly speculation based on Sharman's own attempt at poking fun at the tabloids.

The second non-Soviet woman the USSR took into space was the French Astronaut Claudie Andre'-Deshays Haignere'. In 1996, Claudie flew via *Soyuz TM-24* to the *Mir* space station where she spent two weeks. In January 2001, she flew to the *International Space Station* (ISS) via *Soyuz TM-33*, where she spent 8 days. In my research, I found no evidence or rumors of any intimate affairs on either of her flights. It seems that these flights were all business. However, Claudie's romance and marriage to a fellow astronaut is another story.

French Astronaut Claudie Andre'-Desays Haignere

Claudie met Jean-Pierre Haignere' during cosmonaut training in Star City, Russia. Their ground-based meeting, courtship, and marriage hasn't led to a union in space, however. The duo has never flown in space together, so obviously there has been no possibility of space-based marital relations between them. Both have taken managerial positions, so it seems, for now, there is no possibility of the couple flying to space together on a European Space Agency-sponsored (ESA) launch. However, with the new space tourism industry about to burst onto the scene, the possibility always exists that the duo will become space tourists.

Oddly, only one other female cosmonaut has flown into space. Cosmonaut Elena Kondakova took *Soyuz TM-17* up to the *Mir* Space Station on 4 October 1995 as the flight engineer, and stayed there with her colleagues as part of *Mir* Crew EO-17 until 22 March 1995. This was the first long-duration flight by a team of men and women together. This aroused suspicions that soon became rumors of a sexual space affair.

Orbital Affairs

On the subject of who may have been the first to have sex in space, Space Psychologist Sheryl Bishop told me, "If I had to place my bet, I'd bet on the Russians." She explained, "The Russians are not as inhibited [as Americans] and they did have *Mir* where they had enough room for maneuverability."

Mir Space Station Docked with the Space Shuttle Atlantis

In his article, "The Psychological and Social Effects of Isolation on Earth and in Space," Peter Pesavento suggested that Elena Kondakova and Valeri Polyakov of *Mir* crew EO-17 may have had intimate sexual relations with each other while in orbit. If true, it may have been the first case of adultery in space, since both Kondakova and Polyakov were married to Earthbound spouses at the time.

Rumors of the intimate space rendezvous were reported by the Greek newspaper *Kathimerini* in

Cosmonaut Valeriy Polyakov seen in Mir's window from the Space Shuttle Discovery

Cosmonaut Elena Kondakova – Rumored to have had an affair.

February 1995. The alleged source was a "hint" from Kondakova that "something happened." Valeri Polyakov was in the middle of setting a record for the longest time spent in space on a mission, 437 days, when Kondakova and Alexander Viktorenko joined him for several months to finish the flight.

A possible corroboration is a flirtatious scene caught in a 50th anniversary video from RKK Energia. Onboard the *Mir*, Polyakov is seen squirting water playfully at Kondakova, soaking her while she laughs and attempts to deflect the stream. Of course, this could be innocent fun, but has also been understandably interpreted by many as flirting.

Presumably sometime after the flirtation incident, and for some mysterious reason, Polyakov and Kondakova stopped talking to each other. When asked about it by the media, Polyakov explained, "I had a conflict with Elena Kondakova. It seemed to her that I did not pay enough attention to her in flight. However, I was a bit offended by her disregard for my professional medical recommendations, and this is why I did not feel like communicating with her for a time. But eventually, I realized that it was not right to ignore her and offered to talk about our problem," he said. "She agreed and after this, we became friends again."

In his article, Pesavento says that his deep throat Russian sources were unwilling to comment more on the cause of this period of anger. And Kondakova never did comment publicly on the alleged relationship.

Cosmonaut Valeri Polyakov, record holder for the longest duration space mission – Did he have an affair during his mission?

I don't know about you, but I find Polyakov's statement cryptic. What did he mean, "I didn't pay enough attention to her?" And what in the world were his "medical recommendations?" Obviously, the two had a huge disagreement. Some might even call it a lover's quarrel. The fight, whatever it was about, was significant because the two didn't talk to each other for weeks. In my opinion, when sharing a small, dangerous space station, that's way too long to go without talking.

In Polyakov's defense, he has denied that he had anything more than a professional relationship with Kondakova. Although in one interview he did admit, "It would be desirable to have a normal sexual life in long-term space flights." Wouldn't it though?

Astronaut Alan Bean once made the comment that mixed crews where everyone isn't "participating" could be a problem, "If some are doing it, you are going to want to," said Bean. "Hey, that fella's got a big smile on his face and that bugs me."

Cosmonaut Alexander Viktorenko – Could he have been the third dolphin?

If a romantic or sexual relationship was going on, was the third crewmember, Alexander Viktorenko, aware of it, part of it, or doing anything to stop it? Given the small group and close quarters, it is hard to believe that something like that could be kept a secret for very long. If a relationship did exist between Kondakova and Polyakov, one would think that Viktorenko might understandably begin to show signs of emotional stress. I was not able to find any evidence one way or the other. The Freedom of Information Act doesn't apply outside the United States.

Porn in Space

In the final days of the *Mir*, Russia was looking for new, capitalist ways to obtain funding to allow the space station to remain operational. One idea that was seriously considered involved allowing a production company to send a pair of actors to the space station to film the first pornographic movie in space. Originally titled, "Space Flight has a Price," the name was changed to "The Last Journey." The story follows a cosmonaut who refuses to leave his scheduled-to-be-abandoned space station, so ground controllers come up with the "brilliant idea" to send a woman up to seduce the wayward cosmonaut into returning home to Earth. Oh, yes, a brilliant story line.

The project was led by Russian Producer Yuri Kara who had lined up three actors, Vladimir Steklov, Natalia Gromushkina, and Olga Kabo. The actors actually passed the preliminary cosmonaut physicals and the lead, Steklov, even completed his basic cosmonaut training. The porn stars were in top shape and ready to have sex in space. Yuri Koptev, the General Director of the Russian Space Agency at the time explained, "Life has made us change our mentality and one has to overcome snobbery when dozens of millions of dollars are involved."

Vladimir Steklov during cosmonaut training, almost became the first porn star in space

Sadly, the pornographic *Mir* movie never did happen because the production company could only raise $7 million of the $23 million dollar price that the Russians demanded. The Russian Space Agency wanted the cash first and would not take promises to pay the balance after the film was released. *Mir* was de-orbited and crashed into the ocean in March 2001.

If the film had been made, it would have been the very first porno film shot in space. The porn industry is big business, and no doubt that millions of people worldwide would have paid to see some of the unique ways people can have sex in space. Even without zero-g the DVDs would have flown off the shelves. I think the producers would probably have been able to pay the balance to the Russian Space Agency very quickly, but the agency wasn't willing to make that gamble.

It's too bad that the film wasn't made because it would have been a chance to test those zero-g sexual positions that we'll be pondering in Chapter 2, *How to Make Love in Space*. But someone will make a porn movie in space, probably sooner rather than later. It would cost less to make than most major motion pictures, and would generate millions of dollars, not to mention prestige. We will consider other ways this type of film might be made in the near future, when we talk about space tourism and personal spaceflight business opportunities in Chapter 5, *To Infinity and Beyond: the Future of Sex in Space*. There has been a zero-gravity pornography film made, but we'll get to that a bit later in this chapter.

I'm not *that* kind of Astronaut!

Long duration space flights with crews of men and women have been going on for two decades now, counting all space stations. All told, nearly 40 women so far, have flown into space with their male counterparts. But that's not to say that anything romantic or sexual has ever happened on one of these spaceship and/or space station missions. But there is no proof that it hasn't happened either.

Speaking of long-term space missions, "Of course, it's not like they're just spending a night or two 'in' space," Regina Lynn, Sex Technology Columnist for *Wired* told me. "They're out there for months or years at a time," she explained, "and any time you have a mixed crowd, you have pheromones, even if the initial attraction isn't there."

Rosaly Lopes, Vulcanologist and Principle Scientist at the Jet Propulsion Lab (JPL) explained why we haven't heard about it, "Surely, if people are going to live and work in space, they are going to play in space," she explains. "But, in general, we don't talk openly about our private lives, so I think it is likely that astronauts will do the same."

There is an infamous Internet rumor that NASA has conducted secret space sex experiments. This rumor may date back to a 1989 posting at the University of Iowa. Since then, the story has changed and one version even identifies STS-75 as the 1996 Space Shuttle mission where the sexual experiments were conducted.

It seems that there is no truth to this rumor. Apparently, this is another case of an urban legend that grew out of control. STS-75, which is sometimes mentioned, was an all male crew, so a full-on, *true* test of male-female sexual positions would have been impossible. Of course, astronauts do have a sense of humor. An astronaut may very well have written it, but I found no evidence of that either.

In his March 2000 book, *The Final Mission*, the late French Astronomer Pierre Kohler describes "Document 12-571-3570," which details 10 different male-female zero-g sexual positions, which were supposedly tested by the astronauts. Most of the positions involve physical restraints (such as an inflated tube) to keep the two lovers "docked."

In reference to the documents mentioned in *The Final Mission*, Brian Welch, NASA's Director of Media Services in 2000 said, "We categorically deny there is any such document." The "report" number doesn't sync with the NASA document numbering system, but is associated with the several versions of the original 1989 Internet hoax.

Pierre Kohler's source seems to be a document that has been circulating around the Internet called, "Experiment 8 Postflight Summary: NASA publication 14-307-1792." The document seems to be about space sex experiments conducted on the Shuttle. It's funny to read as it clearly pokes fun at the jargon used in such official NASA documents, but everyone that I consulted about it agrees that it is a fictional document made for humorous purposes. The entire document is included in the appendix of this book, so you can read it later. It's amusing, but the really juicy story about sex in space is still to come.

Shuttle flights are meticulously planned from beginning to end and every experiment is reviewed extensively. Space flights are still relatively uncommon, so there is a great deal of competition to get an experiment flown. Since space and weight on a spacecraft are at a premium, getting a device, an inflatable sex tube for example, onboard for an experiment involves approval by many people, as well as a plan for where to pack it, and when and how to use it in flight. Dr. Rhea Seddon one of NASA's first women astronauts once said of bringing sexual aides and experiments onboard the Shuttle, "I cannot imagine that the many review panels that must approve research on the Shuttle would even let this go forward, that any crew members would sign up for it, or that any Shuttle commander would allow it to be a part of his or her flight," she said.

Officially, there has been no experimentation with sex in space. In 2000, NASA Spokesman Ed Campion said, "We are not, have not, and do not plan to conduct any sex experiments." But what about *unofficially*? NASA has no official *public* policy banning sex between crewmembers, although privately it seems to strongly discourage it. "We depend and rely on the professionalism and good judgment of our astronauts," Campion told reporters. "There is nothing specifically or formally written down about sex in space." This gives us room for some serious sexual speculation.

NASA has no official experiments involving sexual relations, but remember, it has no official policy banning sex between astronauts either. Some in the space community suspect that space sex has probably already happened. In a February 2000 article, James Oberg wrote about sex in space, "Among space flight experts, it is commonly believed that such private activity has actually occurred."

Many people that I spoke with while writing this book also believe that sex in space has already happened. "I don't see why anyone who has been in space wouldn't admit to trying it, if they have. I think it's just as important as any other activity that we're thinking of doing in space," explained Aerospace & Human Factors Engineer Juniper Jairala. "Astronauts have their regularly planned personal time and there's been plenty of co-ed Shuttle missions and space station missions. These people train intensely for months and months before a mission in extreme circumstances and great challenges. That brings people together," she told me. "I'm not sure how they'd pull it off, but I've got to think they have."

Astronauts have been talking about the possibility of sex in space for at least as long as men and women have been blasting off into orbit together. In his recent book *Riding Rockets*, Astronaut Mike Mullane recounts this story about his 1978 astronaut class. "On one occasion," he writes, "someone had pinned up a magazine article on reproduction in zero-g. The author had hypothesized that it would require a threesome to copulate due to the repelling effect of Newton's law… One wit had written 'No! This is why God gave us arms and legs.' Another had tacked up a sign-up list next to the article, requesting volunteers to 'participate in 1-G simulations.' Someone, almost certainly one of the women, had scrawled across it, 'Grow up.'"

It would be difficult for a couple to get a bit of privacy for romance and sex onboard a Shuttle flight, at least not without being docked to a space station, or carrying added crew space. By itself, each Shuttle orbiter has a total crew space of only 2325 cubic feet which is divided into the flight deck, which is filled with seats and instruments, the middeck which contains provisions and stowage facilities for four crew sleep stations, and a small airlock with just barely room for two people. "It's like camping out with your seven best friends in a tent – only you can't go outside," NASA Spokesperson Barbara Schwartz of Johnson Space Center told reporters. "The living quarters are close. There's no privacy."

However, there are missions where astronauts get a little more privacy. Whenever a *Soyuz* capsule or the Shuttle has been docked with the *International Space Station*, or the *Mir* space station in the past, it has allowed more room for crewmembers to spread out, and possibly even find some decent privacy. People could sleep in the Soyuz, Shuttle, or one of the various space station modules.

A Honeymoon in Orbit

While there is no official NASA policy about sex in space, there is an official policy against allowing married astronauts to go on a mission together. This official policy was waived on one occasion when Astronauts Mark Lee and Jan Davis few together in 1992 on Shuttle flight STS-47. This was an unintentional violation on NASA's part. The two astronauts fell in love, got engaged, and then married while training for the mission. Once they were married, mission planners decided not to break up the crew and start over, but to let them fly together.

Space Shuttle *Endeavour* mission STS-47 was flown as a joint life sciences mission between the United States and Japan. In the cargo bay of the Shuttle was Spacelab-J a separate pressurized module with experiments and room for the crew to work. Because of the orbiter's center-of-gravity requirements, the Spacelab cannot be installed at the forward end of the Shuttle's payload bay. Therefore, a pressurized tunnel is provided for equipment and crew transfer between the orbiter's pressurized crew module and Spacelab module. The tunnel is connected to the airlock, and has a jog in it because the doors don't line up. You can't just look down from one end to another. So it is theoretically possible that the two married astronauts, or others onboard, were able to get some private time together in the Spacelab, assuming that everyone else was onboard the orbiter.

Dr. Sheryl Bishop a Space Psychologist at the University of Texas told me, "I have heard rumors about the married couple, that unbeknownst officially to NASA, the crew decided

Jan Davis and Mark Lee in front of Launch Pad 39B with Endeavour in preparation for STS-47

Crew of STS-47 in Spacelab. Jan is at far left, next to husband Mark

that this was absolutely ridiculous and provided some opportunity for the couple to have some off camera time alone." Everyone close to the mission is mum. But I think it is reasonable to assume this rumor may be true.

The STS-47 crew was divided into red and blue teams for around the clock operations. Mark Lee and Jan Davis were on opposite crew rotations, making it difficult for them to be together. Presumably, most of these operations were experiments in the Spacelab, making it a difficult place to be private. It is quite possible that the couple was warned privately by NASA mission managers not to do anything romantic or sexual, perhaps as a condition of allowing them to fly together. The goal for all space missions is to get work done and minimize disruptions. If the crew knew that a couple was busy "getting it on" in the Spacelab, while they were working hard in the orbiter, they would probably be distracted by jealousy, curiosity, embarrassment, or all three.

Star Trek Documentarian Eugene "Rod" Roddenberry put it this way, "I guarantee you it happened, and I have no proof except common sense." And John Spencer, President of the Space Tourism Society expressed it bluntly, "I don't know if there has been sex in space. I suspect there has. Humans are humans, and at least one married couple has flown, and if they didn't do it, they're dummies."

Sheryl Bishop shared with me one reason why the married astronaut couple might not have taken advantage of the situation, "I can see the American couple deciding they aren't going to do that because they didn't want to take the risk. Headquarters would end up grounding them forever," she said.

Officially, NASA has refused to comment on whether it knows if marital relations took place during the 8-day mission. When asked by an *Orlando Sentinel* reporter, NASA spokesperson Kari Fluegel said, "The agency is not interested in the private lives of its employees." Now that's a very non-committal response if I've ever heard one. And it certainly isn't a denial. Is it part of an astronaut's private life if he or she takes time on a space mission to have sex with another crewmember? Is an astronaut's choice of food or music part of their private life too? On a space mission funded by a government, where is the delineation between one's private life and job?

It's reasonable to assume that the official NASA stance is 'whatever happens between two people is nobody's business but their own, as long as it doesn't affect the mission.' Most people would agree that they

Spacelab with crew of STS-47. Notice the ample space separate from the shuttle quarters and floor straps

wouldn't want to tell the world about the details of their private romantic and/or sexual affairs. But on the other hand, there is an intense curiosity because of the extraordinarily unique and exciting location and situation. Sadly, the astronauts themselves are not talking.

When I asked former Astronaut Jan Davis for an interview for *Sex in Space*, her reply was, "No Way. Thanks." She was very helpful and freely talked with me about her passion for space in my first book *Women Astronauts* (Apogee Books, 2002). But on this subject she declined to comment. At least she was courteous enough to send me a reply. Her ex-husband, former Astronaut Mark Lee never did reply to my interview requests. Jan Davis still works for NASA and Mark Lee works in private industry. It may be that the former couple has agreed not to talk about this subject because Jan Davis still works at NASA, an agency that distances itself from sex and romance at every opportunity. Maybe when Jan Davis retires, she will decide to confirm or deny the rumors. But until then, we can put the evidence together and form a hypothesis.

In my opinion, if the astronaut couple did not have sex in space, they are crazy! At the very least they should have had an intimate and romantic cuddle, staring out the window at the Earth and stars.

Shuttle missions are scheduled with bureaucratic precision. Astronauts are told when to sleep, when to eat, when to work, and exactly what to work on. Getting some free time is difficult, and normally cuts into scheduled sleep time. But astronauts do have a bit of scheduled personal time, which they usually spend writing e-mails home or staring out the window at the out-of-this-world view of the Earth and the stars, and, of course, taking pictures. So it's quite possible that a couple could and would spend some intimate personal time together. Most people would have skipped other personal activity to make time to be alone with a loved one and do their own private space experiments. Wouldn't you?

The schedule for work on a space station like the *ISS* is much less densely packed than for a Shuttle mission. Mission planners understand the need for more flexibility over the long months on a station where much of the work involves upkeep and fixing things that break. There is added privacy created by additional space and noise of fans and other equipment. The less ridged schedule, more privacy, combined with the longer duration nature of the mission makes me believe that the space station is a likely environment for a sexual liaison between two people. If it hasn't happened yet, I'd be surprised.

Former NASA flight surgeon Dr. Patricia Sanity said in a 2000 interview with *Maxim*, "If nothing's happened up there, astronauts are less imaginative than I expected." While amusing, she makes a great point.

When the *ISS* was being developed in 1985, NASA Psychologist Dr. Yvonne Clearwater wrote in *Psychology Today*, "It seems obvious, however, that a group of normal healthy professionals will probably possess normal healthy sexual appetites." Clearwater was part of an early NASA space station design team at Ames Research Center that called for soundproofing individual sleeping compartments. Most likely, the soundproofing was to block out general station noise, but I can't help noticing that soundproofing works both ways and provides a more intimate environment necessary to relax alone, with a friend, loved one, or sexual partner. The current *ISS* living quarters are basically closets with a curtain and a sleeping bag tethered to the wall, which, obviously, offers only limited privacy.

Even an unmarried couple or friends may try sex in space because of the novelty. As more women are chosen for long duration missions on the *ISS*, the possibility of sex in space between men and women increases. Of course, the possibility of homosexual relations exists too.

Other Encounters of the Sexual Kind

We've been talking about intimate relations between a man and a woman, but homosexual relations are also a valid possibility. Gay space sex may very well have happened, especially on a space station.

Space Activist Derek Shannon told me why he believes that both gay and straight sex have happened in space, "The best evidence I've seen for sex in space is Cosmonaut Sergei Krikalev, as featured in the IMAX Space Station: 3D. He has more cumulative time in space than anyone else, and I can't imagine anyone, male or female, being put in a tin can with him for more than few weeks without making a pass. The man is hot! Now, whether Sergei has accepted any propositions is between him and his lovely wife Yelena, but I would hope that a mutual understanding is not out of the question," said Shannon. "There must be some reason Sergei keeps getting picked for all these long duration flights!"

There are some good reasons, however, why homosexual relations may not have happened yet. A majority of astronauts have been chosen from the ranks of the military, and all of the pilots and mission commanders are military pilots. This is true both in Russia and the United States. In both countries, there is an unfortunate intolerance of gays in the military. This is not to say that there are no gays in the military, but there are probably less than in the average population. Add to that the fact that gays in the military must hide their sexual preferences. This makes a larger number of military personnel even more homophobic than the average population seems to be. When a military member becomes an astronaut or cosmonaut, he or she is still in the military while doing their job as an astronaut. So if he or she were gay, they would still not be at liberty to reveal it and discuss it openly with other astronauts and cosmonauts.

"Hot looking" Cosmonaut
Sergei Krikalev

With so few women in the astronaut program, the opportunity for a lesbian relationship between two women astronauts on a mission seems even more remote. There have been a number of Shuttle missions with two or more women, but no long-duration station missions, either to the *ISS*, *Mir*, or any earlier orbital station.

One comment in former Astronaut Mike Mullane's book, *Riding Rockets*, is especially interesting. When discussing the astronaut class of 1978, he writes about the diversity of the group and how there were several non-military civilians in the group as well as women for the first time. He writes, "Truth be known, there were probably gay astronauts among us." Is Mullane speculating? Or does he know something, and decided not to reveal it out of respect to a friend or colleague in the group who is still "in the closet?"

Homosexual sex in space seems less likely to have happened than heterosexual sex in space. NASA, in fear for its public funding, doesn't want any segment of the public to object to any action it takes, or people it chooses to become astronauts. It seems ridiculous, of course, that sexual orientation should be a factor in choosing the right person for a job in space, or for any other job. The reason for this is the number of homophobic people in the United States is a significant segment of the population. So a gay astronaut would need to hide his or her sexual persuasion not only from the public, but from NASA selection boards, NASA administrators, colleagues, and fellow mission crewmembers.

Public knowledge of a homosexual astronaut might be career ending, at least in the past, and probably now too. For the same reasons, having sex in space with a gay partner on a mission with other crewmembers nearby would be nearly impossible and almost certainly career ending unless all other crew members were intimately trusted. Maybe someday soon we will see this attitude change.

Finding another gay astronaut on a mission would be difficult given that there are probably fewer gay astronauts than in the general population. Finding another gay astronaut, and being sexually attracted and interested in each other are two very different things. But on a long duration mission, a person (gay or straight) may not be as picky as they might be on Earth. We will get into that in Chapter 4, *Sex on the Brain and Lust in Space*. As a bonus, the selection process and basic training assures that all astronauts are in top physical shape. Hopefully someday soon, a gay astronaut will tell the world that he or she is gay and will prove to the world that sexual orientation has nothing to do with job performance. This probably wouldn't happen, however, until the astronaut retired or was convinced that his or her days of flying into space had ended.

It's quite possible that gay sex has happened on one of the many two-man space station missions, but I found no evidence or rumors of this. These missions last for six months or more, usually with one American and one Russian. If it has happened, then there would be great pressure to hide the relationship in order to prevent damaging a career or marriage. It would be a huge risk, and would probably never be discussed. On the other hand, astronauts are in a risky business. It's possible that this translates into taking sexual chances, including gay sex, threesomes, and group orgies in zero-g.

Flying Solo

What about solo sex? Unlike sex between partners, it's almost a certainty that both male and female space travelers have masturbated in space. There have been a lot of people, men and women, who have been on long-duration space missions, sometimes for months at a time. Even if some people decide not to masturbate, most would. Even on a short-duration flight of a few days, some would feel free to indulge on their off duty time. That is one of the personal choices that NASA says that it has no interest in learning about. However, it is interesting to note the loss of muscle mass in the lower body on a long-duration space mission, may reduce sexual stamina in zero-g.

Cosmonaut Valeri Polyakov said that towards the end of his long duration stay, "Psychological Support Service sent us some nice 'colorful' movies which helped

Skylab in orbit at end of mission

to recover our will, to act like a normal adult male. There is nothing to be ashamed of." I interpret 'acting like a normal adult male' as masturbating after, or more likely while, watching porn. At least I think that's how Polyakov meant it.

In his book *Liftoff*, former NASA Astronaut Michael Collins explained that prior to launching the *Skylab* space station, there was a concern by the medical group that complete celibacy could lead to infected prostate glands and urinary tract infections. So, as Collins put it, "One doctor advised regular masturbation, advice Joe ignored." While he was talking about *Skylab* astronaut Joe Kerwin in that paragraph, the implication is that others, including Collins, did not ignore that advice and did pleasure themselves.

However, on the next page of his book, after mentioning the lack of booze on *Skylab*, Collins says, "There was no sex on *Skylab* either." He goes on to say, "Because of the heavy workload and preoccupation with other matters, the absence of females wasn't a big problem for any of them." Astronaut Alan Bean once likened it to being on a Navy ship, "No sense thinking about it," he said. The impression is that he was clearly talking about sex with a partner in this context because, in the very next paragraph, he goes on to speak about Antarctic missions with men and women. So, it is easy to conclude that masturbation was a personal decision and, just as here on Earth, some indulged while others chose to abstain.

Getting Wild

We know that Earth orbit has not been entirely celibate. There have been documented and filmed cases of space sex! Well, sex between *animals* anyway.

NASA biologists sent a number of fish to the *Skylab* space station hoping they would mate and have babies. The fish went berserk in weightlessness, looping frantically and were never able to adapt. Unfortunately, there is no possibility that they had sex or reproduced.

In 1979, the USSR wanted to be the first to document sex in space. So they sent up five female rats and two male rats to do the deed. After 19 days in space together, however, there was no evidence at all as to whether the animals mated or not while in orbit, and none of the females gave birth after returning to Earth. Maybe the rats had performance anxiety.

In November 1985, the last successful flight of the Space Shuttle *Challenger* included 240 female and 90 male fruit flies. The flies bred in space, laying and fertilizing eggs. Many of the embryos failed to develop into larva, but this appears to be the first documented successful breeding of animals in space.

Jan Davis and Mark Lee check latches on rack 5, the adult frog compartment on STS-47

In 1992 on STS-47 *Endeavour*, the same Shuttle flight with married Astronauts Mark Lee and Jan Davis, four female South African Clawed Frogs were injected with hormones on orbit to induce ovulation. Then their eggs were saturated with frog sperm to see if any would hatch into tadpoles. They did! This was the first instance of non-insect reproduction in space, albeit with artificial insemination.

Both the Russian and American space programs sent various vertebrate animals such as mice and rats into space. Typically, these poor little animals were frightened by weightlessness and clung to cage walls or other surfaces desperately seeking normality. But in at least one case, on the *Cosmos 1129* satellite, male and female rats did mate. No pregnancies resulted, but postflight laboratory tests showed that ovulation and fertilization did occur, but for some reason the embryo didn't form.

This led to the search for a vertebrate animal that would adapt to weightlessness, and hopefully, breed. Animals were taken on aircraft and subjected to weightlessness and testing. Reportedly, this was as big and as intense as the hunt for the first human U.S. astronauts.

Finally, a Japanese team led by Kenichi Ijiri of the University of Tokyo, managed to find a strain of fish that didn't go berserk when weightless, at least in short ballistic arcs flown by aircraft. Instead the red-orange fresh water fish, called Medaka, would orient themselves based on a light that was shone into their tanks to mimic the light of the rising Sun. Ah, poetic fish.

NASA sent two pairs of these fish into orbit in 1994 onboard the Space Shuttle *Columbia*. According to the Washington Post, "Ijiri noted that the two Medaka couples, though experienced lovers, failed at initial attempts in orbit. Also, he discovered, the fish set a record for 'jealous' pecking attacks on each other, with a female sometimes attacking a couple during mating."

After three days, the Medaka delighted the team watching them by mating during a live video downlink. "I could see with my own eyes this dramatic scene," said Ijiri. This was the first documented mating in space of vertebrae animals. This video is what former NASA Chief Scientist Dr. Kathie Olson jokingly referred to in her lectures as the "Skin Flick."

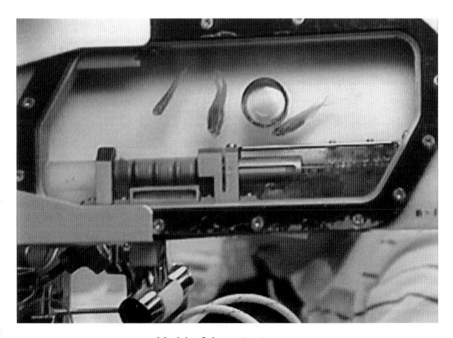

Medaka fish mating in space

The biological process worked! The two pairs of Medaka fish gave birth to eight healthy offspring in space. The four fish became grandparents and then great-grandparents when their offspring had babies back on Earth.

Medaka fish born in space with his mother

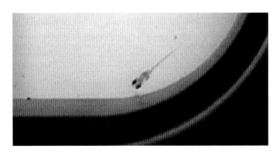

Closeup of baby Medaka fish (5 mm length)

Sex Experiments in Zero-g

Yes, there has been sex in zero gravity here "on" Earth. You can simulate the zero gravity conditions of space in a couple of different ways. First, I'll explain how to achieve weightlessness on Earth, and then we'll get to the sex.

Weightlessness can be achieved by falling. If you don't have anything pushing you up, like the ground or air friction, then you are in weightlessness, or "experiencing zero-g." Skydiving is similar to zero-g because you are falling, but very quickly the human body reaches its "terminal velocity" and the friction from the air pushes against you, slowing you down. Falling in this way is close to zero-g but is still not the same feeling. We need something more realistic for our passionate purposes.

Flying an airplane in ballistic arcs can create a more accurate simulation of zero-g. Near-weightlessness is produced inside the aircraft for a short time as the airplane goes into freefall. The idea of putting an airplane into freefall on purpose sounds scary, but it truly is an art and a science. A "ballistic arc" is simply the path that the airplane must fly to produce zero-g. The aircraft starts with upward momentum and then follows a flight path that matches freefall. So the path is similar to the path of a rock you throw up into the air at an angle. The result is weightlessness inside the aircraft for a short time, followed by a period where the aircraft "pulls up," which means it generates lift and flies upwards to avoid the ground and start another ballistic arc. This is when you get more than 1-G, higher than normal Earth gravity, also called hypergravity. This continues until the aircraft is ready to repeat the freefall path.

The advantage of the ballistic flight is that you are inside the airplane and do not encounter wind resistance as you would if skydiving. And you don't need to wear a suit and parachute. In fact, you could wear nothing at all. An interesting side effect of the ballistic flight is that you must endure higher than normal gravity in order to experience zero gravity.

The amount of time that you have to fool around in weightless conditions depends on the initial velocity and angle, and the capability of the aircraft to pull out, or "up" from the maneuver. Typically you'd have about 20 seconds of zero-g flight. It's not much, but it's the best we can achieve here inside the Earth's atmosphere. NASA and the U.S. Air Force use this simulation to test experiments for spaceflight and for astronaut training. Moviemakers Ron Howard and Tom Hanks used it to create the realistic zero-g shots for the 1995 movie *Apollo 13*. Today, the Zero-G Corporation offers commercial flights to anyone who wants to experience zero and high-g conditions, as well as the gravity of the Moon and Mars. Now, onto the sexy stuff.

Floating in weightlessness with aircraft ballistic arc flight path illustrated.
Courtesy of Space Adventures

The Spanish-produced sex video, *The Uranus Experiment 2* includes a scene filmed in weightless flight. One particular segment is supposed to depict several male and female astronauts having sex in weightlessness, but was obviously filmed here on Earth, and no amount of rotating camera angles can fool the viewer into believing that this is a weightless shot. But at the end of the scene, the 20-second "money shot" is indeed filmed in weightless conditions on an aircraft. However, it's not what former U.S. President Bill Clinton would refer to as "sex." Unfortunately, no sexual positions, like the ones you will soon read about in this book, were tested. Clearly this part of the movie was filmed in weightlessness, and just as clearly, the producers wasted an opportunity to show or demonstrate anything unique and exciting. What a missed opportunity! Besides, the guy got all the action. *How unfair.*

Another way to simulate a zero-g spaceflight is in the water with neutral buoyancy. Neutral buoyancy is achieved when the buoyancy of the body is exactly matched by the weights of the suit being worn. Astronauts train for space walks in large neutral buoyancy pools at Johnson Space Center and Marshall Space Flight Center. The advantage of neutral buoyancy is that it can be sustained for long periods of time. The disadvantage is that you have to be wearing a big, cumbersome spacesuit to achieve neutral buoyancy.

Astronaut Susan Helms trains for EVA underwater in the
Marshal Space Flight Center Neutral Buoyancy tank

In his book, *Living in Space* the late NASA Consultant and Author G. Harry Stine claimed that the neutral buoyancy tanks created for astronaut training at the Marshall Space Flight Center were used to test whether people could perform sexual "docking maneuvers" without the effects of gravity. According to Stine, the sex was, "possible but difficult." He figured that sex without gravity would be easier if someone were holding the two participants in place.

Stine wrote, "It was nice to have somebody else there to push at the right time and the right place. And then when they got done with the experiment, lo and behold, they discovered that this is the way dolphins do it. There's always a third dolphin there to push at the right time and in the right place," he wrote. Well, the truth is, that sometimes dolphins assist each other and sometimes they don't. A wonderful 1997 documentary called "Dolphins with Robin Williams," shows dolphins mating. In this skin flick, there was no "third dolphin" helping out. Sex was strictly a fast wam-bam-thank-you-dolphin encounter. Literally, if you blinked, you missed the action.

Even if the "three-dolphin" experiments were performed, and I have serious doubts, it certainly wasn't an experiment performed officially by NASA. Such sexual position experiments would have to have been done unofficially by neutral buoyancy tank personnel and some very special volunteers.

Most people, but not all, would be extremely uncomfortable with the idea of a third person holding them in place while they have sex. Some sort of device to hold two sex partners in place would probably go over much better with most people than the group 'project' suggested by Stine. But for some, three is not a crowd.

In the next chapter, we will delve into more of the details, and some device ideas, that may help us to figure out how to make love in space.

Chapter 2: How to Make Love in Space

Mars and Venus in the Space Bedroom

When you think about it, sex in space is space tourism's "Killer App." As space tourism becomes a reality, singles, couples, and trios alike, will venture to hotels in low Earth orbit expecting to try the ultimate space experience, zero-g sex! Physiologically, there is every reason to believe that both men and women can reach orgasm in space, so you're probably very interested in trying zero-g sex sometime in the not too distant future. Are you curious? What do you expect it to be like?

In this chapter, I will take what we know about the space environment, and what we know about sex on Earth and consider some ways you might make love in space. Someday, a *Cosmic Kama Sutra* will guide the new space traveler into sex on the final frontier. It will, no doubt, be essential tour book reading at your local spaceport. This chapter may be the first cut at that book. It's what everyone who intends to have sex in space should know, is afraid to ask, or didn't know whom to ask. One thing is certain, for those willing to try, some fun experimentation is ahead.

What will your living conditions be like on this future space vacation? Obviously living and loving in space is very different than here on Earth. But what exactly is different? How do these differences affect sexual pleasure? Our first question vis-à-vis the possibility of sex in space is; what is the space environment?

Falling in Love with Weightlessness

The biggest and most obvious difference between Earth and Space is the lack of gravity. Experts talk about micro-gravity, which is essentially very small gravity caused by the masses of the space station or hotel itself on the experiments or, in our case, people. This level of gravity is undetectable except by precise instruments, so for our purposes – sex, it is the same as no gravity, also known as *zero-g* or *weightlessness*.

In orbit, the Earth's gravity does pull on you. If there were no gravity in orbit, you would go flying off into the void of space. That might not be as much fun as it sounds, but I digress. The small amount of Earth gravity pulling on each person is the same as the amount of gravity pulling on the spacecraft, station, or hotel, so there is no *feeling* of gravity or weight acting on your body. For all intents and purposes, and you know what purposes those are, you are weightless. It's faster and easier than any weight loss plan on Earth. When in orbit, the hotel and everyone inside are actually *falling* towards Earth.

ISS crews eat together in space. Notice the weightless tortilla. Courtesy IMAX

So, we've established that the Earth's gravity does pull on people and things in Earth orbit, but since everything is pulled in the same way, you feel weightless. The feeling of weightlessness is like falling, without the wind resistance that you would experience while falling from an airplane or tall building, and without the unfortunate splat. On Earth, you experience something similar when floating in the water. Your buoyancy *counteracts* gravity in the water. Weightlessness is kind of like being in the water, only more extreme, and with more possibilities for fun of all kinds.

Like a swimmer, when you are weightless in space, any push can send you moving in one direction or another. But in space, there isn't the resistance of water to hold you in place, so only the slightest nudge is needed to move. The air in the cabin will resist your motion a little bit, but not nearly as much as water. It usually takes astronauts a few hours to learn how to move around well enough to work in weightlessness. Even then, it's a very different way to live, work, and play, especially considering that you have lived your life in the 1-gravity environment of Earth.

Astronaut Wendy Lawrence told me that she loved, "Being able to fly around like a bird. It's just very fun to not feel the effects of gravity, to just, no kidding, be able to move yourself out to the middle of the cabin and let go and completely relax, and be perfectly suspended right there in the middle of the air," she explained. "It's just fun."

Weightlessness can be disorienting and it takes a little time to get used to the odd feeling. Astronauts use a mix of handholds, footholds, pushing, and floating to move around inside the cabin. In order to stay in one spot, you normally have to hold yourself down, and that takes work. Aerospace Medicine expert Dr. Harvey Wichman explained how this is tiring to astronauts, "A large portion of the time, an astronaut is holding on with one hand while trying to work with the other hand. Then a good portion of the time [the astronaut spends] finding unofficial ways to restrain themselves by jamming a foot into or between something that wasn't intended to be a restraint. And then every now and then they used the foot loops."

What surprises some astronauts is the rotational motion they get when pushing off a wall or steady object. People don't push off perfectly straight, so each time they get a little spin. Some astronauts feel like everything is upside down and that they are hanging in space, which is basically true. But, once astronauts get used to it, weightlessness is fun. Imagine spinning in place, and performing feats that would be considered amazing here on Earth. On almost every mission, astronauts have made videos of themselves floating in midair, spinning in place, and making their food, pens, and anything else they can think of into zero gravity toys. Perhaps they have also used each other as toys.

"Astronauts tend to have a really good time in weightlessness, and that was true during both of my flights," former Astronaut Sally Ride told me. "During the off hours, we'd basically have a good time doing somersaults in the middle of the cabin or floating peanuts from one side of the room to the other."

Imagine the possibilities for playing with your significant other. It might be fun to watch a squirt of whipped cream fly across the room. But don't squirt the water! Harvey Wichman explains, "You cannot have a liquid that is loose in space. Everything in space has to have a fan to cool it because convection currents don't occur in space. Warm air doesn't rise. So every piece of apparatus has fans sucking air in and blowing across the electronics. You get a little bit of liquid free, and it's going to get sucked into electronics and short everything out," he says. We'll talk more about this in *Chapter 5: To Infinity and Beyond: the Future of Sex in Space*, when we look at the ideal honeymoon suite and how you might handle those interesting spills.

On Earth, you can put things down, but in space you can't. Astronauts tend to put things in the air for short periods of time and then grab them when needed. In videos taken in space, you can see astronauts place a toothbrush in the air, then reach for the toothpaste and a squirt bottle, picking up the toothbrush again later. You can't do that on Earth. Objects in space, unless fastened, don't stay in one place for very long. Air currents and slight shifts in a spacecraft, station, or hotel will cause objects to move off slowly. So if you shed your clothing it would drift around with you for a while until air currents made it float away.

Astronaut Shannon Lucid once described to me how on the day after getting back from a long duration flight on the Russian space station *Mir*, she put a cup in mid-air, let go, and was briefly surprised when it dropped straight to the ground. She was so used to objects staying where they were placed in mid-air, that she momentarily forgot that things don't work that way back here on Earth.

A swimming pool is a good place to simulate weightlessness, and zero-g sex. Astronauts practice their space walks (Extra Vehicular Activity or EVA) in a large, specialized pool. And as we learned in Chapter One, there are some rumors about sexual experimentation with the "three-dolphin club." While sex in a swimming pool is much cheaper and more accessible than sex in space, it is not a great simulation and has some shortcomings.

Astronauts training underwater in neutral buoyancy chamber with divers

To truly simulate weightlessness you must use weight belts or something similar to achieve neutral buoyancy. Too little and you float, too much and you sink. You will know that you have achieved neutral buoyancy when you are floating underwater, neither sinking nor floating.

The biggest difficulty, obviously, is breathing underwater. If the activity is strenuous, as sex should be when done right, holding your breath becomes nearly impossible. A snorkel or scuba gear can help fix this problem, if it doesn't get in your way. Another difficulty is that water tends to carry away, shall we say natural (or artificial) lubricating sexual fluids, which could cause problems.

With neutral buoyancy and a way to breathe, you could theoretically have a great time practicing simulated space sex. Be sure to use arms and legs to hold on to each other. But remember that the water provides resistance that you wouldn't have in space. So it's not the same as sex in space, but could still be fun.

Having sex in the weightlessness of space, you would avoid the difficulties of breathing and water resistance, and it would most likely be much more enjoyable. No breathing equipment or weight belts to get in the way of kissing and hugging. You would still have the difficulty of staying in place, and all the fun of having to hold on to each other. For all of these reasons, I would think that sex in zero gravity, or even in low gravity, would be fun.

"I imagine it as being similar to sex under water but with noticeably less resistance," *Wired* Sex Technology Columnist Regina Lynn told me. "It would be a challenge to touch without pushing each other away. You'd have to hold on, make sure you exert gentle force to keep bodies together and not push apart. Having no up/down, left/right orientation would make your bodies the sole reference point, unless of course you're in a chamber with handholds and padded walls and clearly marked hatches," she explained. "Zero-g sex sounds incredibly intimate to me."

Astronaut Gerald Carr pretends to balance Astronaut William Pogue in weightlessness on *Skylab*

True Space Sex

Astronaut Edward Gibson did his own little test on *Skylab* to see how long he could float, starting in the middle of the room, before air currents drifted him into a wall. It took about 20 minutes. In a space hotel suite of sufficient size, this would be the most unique form of sex. Perhaps this is the truest form of sex in space.

Imagine a time several years from now when you have become a space tourist. You've been in orbit for a couple of days and have adjusted to weightlessness. Now you are ready for a bit of romance with your partner and choose some music to play in the background. You may want to try cuddling, dancing, or something more energetic.

You float in the middle of your own private space hotel suite, using your arms and legs to hold on to each other and nothing else. You might look "up" through the window at the breathtaking Earth and brilliant stars, and back into the eyes of your very significant other. Personally, I think that floating like this would be the most unique and romantic experience in the universe. Having sex while floating this way should officially be called *true space sex.*

So how do you get to the middle of your space hotel suite with your partner? It's not as easy as it sounds. Author and Space Activist Vanna Bonta shared with me her romantic challenges in zero-g while on a ballistic flight. "Having personally *kissed* in zero gravity," she says, "I was initially amazed by the unexpected lack of attraction, from the sheer perspective of the mass magnetism. We wanted to kiss, that attraction was present, but as our faces and bodies drew nearer, gravity was not helping us at all," she explained. "It took minor struggle to come together, and we had to hold on in an effort to keep our lips aligned."

If you "jumped" towards the middle of the room, you would keep going past the center and past your partner, unable to stop. You would go splat into the wall on the opposite side of the room. Fortunately, there are two other basic methods that would work.

The first is to gently jump to the middle of the room while holding onto something (the "heavier" the better). Then, when you reached the center of the room, you would push the object you are holding ahead of you towards the wall that you are trying so hard not to fly into. This kills your momentum and lets you stop in the middle of the room. Of course, the object you push away may come bouncing back unless it is covered in something sticky, like Velcro, and will attach securely to the far wall. By the way, Velcro is severely limited in space because in the event of a fire, the fumes from burning Velcro are toxic. So we must use only small amounts of Velcro until we can find an alternative material to make Velcro out of.

The second technique to stop yourself in mid-air takes a partner, but this shouldn't be a problem. You and your companion start on opposite sides of the room and jump towards each other. When you meet in the middle, you must grab on and hold tight. If you do it right, you will cancel momentum with each other. If you miss, you can still have fun spinning off of each other and slowly moving back to the side to try again.

Whether you are floating in the middle of a room, off to a wall (all sides including "up" and "down" are "walls"), or bouncing around a room together from wall to wall, the weightlessness of space offers unique and exciting opportunities for couples. Besides, it just sounds like a whole lot of fun.

Those who've experienced the "squished arm syndrome" while cuddling in bed will love cuddling in space. This happens when a couple is hugging in bed the lowest arm gets, well, squished. This is especially true if the lower arm is around the waist or chest, instead of the neck where there is a convenient tunnel for one arm. In space, without gravity, squished arm syndrome can't happen. Yay! Now you can cuddle for hours any way you like and avoid the annoying problem of body parts falling asleep because they were squished between a body and a bed. Long live the zero-g bed!

Space Sickness

Following launch and "orbit insertion," the transition to the weightlessness of space confuses the human body and can lead to Space Adaptation Sickness (SAS). About half of the astronauts feel sick when they first arrive in orbit, and it's not necessarily the ones who are prone to seasickness here on Earth. Unfortunately, there is no way to tell which astronauts will be affected.

"I vomited my guts out," says Millie Hughes-Fulford of her 1991 Space Shuttle flight. As a payload specialist, Hughes-Fulford had gone through nine years of rigorous flight training with her crewmembers. "I didn't get sick the whole time before," she said. Luckily, a shot of anti-vomiting medication helped to take away her nausea.

Phenergan is the drug most often used by astronauts to reduce the effects of SAS, and is injected for fast action. Unfortunately, it also makes you drowsy and out of it, so you're not doing much of anything while under its influence.

Automatic injectors for pain (Demerol) and motion sickness (Tigan) as used on Mercury-Atlas 9

Astronaut Anna Fisher also had a bout of Space Adaptation Sickness the first few days of her first mission. "I remember thinking that I don't ever want to come back," she said. Thankfully, medicine helped in her case too. "By the fourth day it was 100% gone and it was so much fun," she explains. "And I remember wondering why I felt that way. But then, I used to get seasick when I went scuba diving, so I learned to just deal with it."

Astronaut and Medical Doctor Ellen Baker says that some of the astronauts, "feel a little queasy and maybe even vomit the first two to three days of the mission. But," she explains, "that generally passes and we actually have some very good medicine for them." Typically, by the third day of the mission everyone has recovered from SAS and is ready to work and experience the fun of weightlessness.

The feeling of vertigo and falling, combined with confused nerves, can make an astronaut feel space sick. Without gravity there is no "down" or "up" and the inner ear, which controls balance, becomes confused. Then, when the astronaut moves his or her head, the fluid gets sloshed around. The symptoms are similar to carsickness, airsickness, or seasickness - so you would think the causes are the same, but based on research there appears to be no correlation between motion sickness on Earth and Space Adaptation Sickness.

You have to teach people to vomit," says Harvey Wichman. "NASA has very special vomit bags. You cannot have loose vomit running around. Not only is it wet, but it's corrosive. It's mostly hydrochloric acid. What doesn't get shorted out will get eaten away in no time." The bags are held easily in one hand with a cone and an absorbent layer in the bottom, like a baby's diaper, to prevent splashing and absorb odor and moisture.

In all cases, SAS disappears within a few days. Half of all space travelers get SAS, so as a precaution, tourists should plan to stay in space for at least a week. This way, they can be sure to have some fun after the adjustment phase.

Vanna Bonta explained why space sickness might not be as bad as you think, "On Earth, much of the wrenching discomfort of emesis, apart from the sensation of nausea itself, is from the coordination of many muscles it takes to counter gravity. Some animals on Earth regurgitate as opposed to vomit, i.e. stomach contents flow up into the esophagus without any forceful abdominal contractions. What I experienced in zero gravity was similar to this, expulsion without the heaves. As long as you've got the mouthwash with you, Space Adaptation Syndrome (another term for throwing up) doesn't have to ruin the date or detract from romance."

Upon arriving at your space hotel, you will probably want to give yourself a day or two to acclimatize to the new environment before trying anything bold and

View from space of the eruption of the
Cleveland Volcano in the Aleutian Islands

new, like space sex. SAS might make at least one of you not in the mood. Even if you weren't sick, the exercise might induce space sickness where there was none before. New space travelers should be advised to relax and enjoy the view and get used to weightlessness with unique cuddling. Just hold onto each other while you float and look out the window at the out-of-this-world view.

Changes to Your Celestial Body

When on Earth, gravity holds down the bulk of bodily fluids such as blood. When gravity disappears suddenly, as it does after launching into space, the head receives an excess of these fluids. In weightlessness, the blood and other fluids that are usually dragged down towards the lower body by gravity, are released and tend to redistribute themselves and "rise" towards the head. This can lead to light-headedness, clogged sinuses, a stuffy nose, and can feel like a bad head cold for the first day or two until your body gets used to the fluid shift. This makes your face get puffy. Although I don't recommend it, the best simulation would be to lie down on your back with your feet propped up a bit higher than your head. Hold that position for 30 minutes then look in a mirror. Your face should look puffier and you will probably feel stuffed up. This is one of the things to expect as your body transitions to the weightless environment of space.

The good news is that the extra fluid in your face will tend to make wrinkles go away, if you have any. So, you and your companion may look younger and thus feel more attractive, and maybe even a bit more sexually excited about one another. This is good news for those hoping to create spas in orbit. The anti-wrinkle effect of space may seem like the fountain of youth. Unfortunately, when you return home, the wrinkles return with you. It must be nice while it lasts though.

The worst part of being in space for Wendy Lawrence was, "The adjustment from a gravity environment to an environment where you don't feel the effects of gravity," she explained. "Your body has to go through an adjustment period. So for me that's about 24 hours, and my inner ear doesn't like that the gravity has been removed. So you just have to be kind of careful when you go through that adjustment phase. The fluid has shifted to your head, you have a headache, your nose is stuffed up, and your inner ear is confused. Once you get past that period, it's very comfortable to be in space," says Wendy. "That's a small price to pay."

During the first several days in space the body dumps what it considers to be excess fluid, mostly water. This means you will need to use the bathroom a lot, so make sure to empty your bladder before starting anything romantic or energy-intensive, or else you may find yourself interrupted.

Unfortunately for those spending more than several weeks in orbit, the body starts to lose muscle tissue and bone density. This is because the muscles and bones are not being stressed constantly by gravity as they are on Earth. The heart becomes weaker too because it does not have to work as hard to pump blood in weightlessness.

However, it takes time for the body to lose significant muscle and bone, so a couple of weeks in orbit will not cause any long-term debilitation. But if in orbit for months, the lack of stress on muscles can have severe effects upon return to Earth. Appropriate space exercise will help preserve muscle tissue somewhat. Typically this means that space travelers will have exercise for an hour or two each day. Studies suggest that it is best to break this into 20 or 30-minute sessions. That takes up a good part of the day, making long duration missions less efficient.

Sex in space may be the ideal exercise to prevent muscle atrophy. During sex, the heart beats faster, muscles in the buttocks, abdomen, chest, and even the face, hands, and feet, contract and spasm. The act of holding on to each other in zero gravity would take a lot of energy too. The resulting vigorous flexing and muscle spasms make space sex a better exercise than stationary bicycle riding or any of the other exercises that astronauts and cosmonauts do now to keep in shape. But you would have to be diligent and have sex every day, maybe even twice a day to gain the benefits.

Studies have found that having sex regularly can keep a person healthy because the activity somehow helps the immune system stay strong. This will be especially important in space where zero-g suppresses immune function. Sex may truly be the way to counteract many health problems. This is why it is so important that the space agencies study sex-related issues.

The loss of bone density that occurs during long duration space voyages is almost impossible to prevent. There are some drugs, used to treat osteoporosis on Earth, that may help in space. But the only real solution is to give your skeleton the chance to hold up your own weight. This can be achieved by spending a certain amount of time on a planet, moon, or in a centrifuge.

This is another area where the space agencies need to do some important research. We need to find out exactly how much, or how little, gravity the human body needs in order to keep bones strong on a long trip into space. This won't be a significant concern for space tourists on a three-week vacation in space. But as vacation times get longer, and we eventually take long vacations to the Moon and Mars, bone loss will become a serious issue.

It cannot be seen or felt, so you might not notice overexposure until it is too late, but radiation exposure is very dangerous. You wouldn't feel it, but radiation is affecting your body, even now while you are on Earth. Radiation has an effect on the body's reaction, and perhaps more importantly, it has a large affect on any potential procreation in space or afterwards when on Earth. We will cover the effects of radiation in more detail in Chapter 3, *Making Space Babies*.

Basically, higher levels of radiation hit your body while in space. Astronauts and vacationers to the Moon and Mars will be exposed to high levels of radiation. Even in Earth orbit, a traveler no longer has much of the Earth's protective blanket of atmosphere to shield them from radioactive particles flying around in space. However, if a tourist stays in low Earth orbit, rather than going to the Moon, or further out into space, the Earth's magnetosphere will still be quite effective at keeping back a lot of the more dangerous, highly energetic, charged particles in space. So while radiation levels are much higher in low Earth orbit than on the Earth's surface, they are still lower than the radiation levels endured by a space traveler journeying beyond low Earth orbit. Earth orbit is kind of a safety zone for tourists, it seems.

The Space Spa

Don't worry, not all of the changes due to weightlessness are bad. When in space you grow a couple of inches taller as your spine unbends a bit and the compression between discs is relieved. Your waist and legs become thinner as they lose fluid that is normally forced downward. Astronauts have reported as much as a three-inch reduction in waist circumference within a few days of getting to space as the fluid leaves that area. Your face becomes puffier because the fluid that is normally drawn away by gravity moves to the head, making your cheeks more full and smoothing facial wrinkles. Your body tends to loose pounds, but mostly from fluid. Chests gain fluid, so female astronauts may see an increase in breast size. And without gravity to pull them down, breasts seem to defy gravity, and look more 'perky.' Everyone will agree that this is a nice adaptation. The combined effect may make you and your partner look a little younger, and who doesn't want that?

Former Astronaut Susan Helms liked how her body changed while she was on the *International Space Station* for six months, "I got taller. I shed about 20 pounds. Your legs get very skinny. If you have varicose veins, they'll go away. Wrinkles go away because the fluid shifts. You're getting the idea that they could build a spa in space and there would be a lot of people paying money to go," she jokes. "It's almost like a fountain of youth in a sense. Your body goes back 20 years and sheds the effects of gravity. It's pretty amazing," she says. Unfortunately, it all reverses in a short time once you return to Earth.

A very young looking Susan Helms in space as photographed on STS-54 in January 1993

Increased fluid in the face causes some senses, like taste and smell, to weaken a bit, at least for a while after arriving in space. Astronauts tend to like spicier foods while in orbit because this congestion prevents all but the strongest smells and tastes from being noticed. Senses always return to normal upon return to the Earth, so there is nothing to worry about here. Hotels may want to spice up their menus for spicy-minded clientele, however.

Other senses, like touch, could be enhanced simply because you are not in contact with anything else. This would work in the same way that sensory deprivation enhances senses. But it is important to note that prolonged "exposure" can lead to illusions. Gene Myers, CEO of the Space Island Group, a company working to finance a private space hotel for tourists, once said, "In weightlessness, if neither partner is wearing any clothing, the only physical contact is the touch of a partner, a kiss, a caress. The physical experience and sensations of contact would be much more intense than anything you could get on Earth," he said. Imagine the possibilities.

After a day or two of adjustment to the space environment, your stomach is under control and you start to notice that your partner looks even better than usual. You look great too! You're both a bit excited, so maybe it's time to leave the lounge area and go to your own space hotel suite together. What can you expect? Let's find out.

Intimate Space Hotel Accommodations

Caution. While staying in space you should always sleep strapped to something. Just like objects, people who are asleep tend to float around at random and can hit things, which can be dangerous to both the person and the hotel. This doesn't mean you need to be strapped down tightly, it just means that your sleeping bag needs to be on a short leash.

When floating in space, you don't have gravity to press your body to the bed, so your limbs float freely and there is no risk of them becoming numb by sleeping in the wrong position.

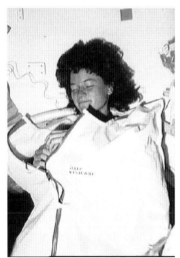

Astronaut Sally Ride asleep in a sleep restraint (sleeping bag)

Astronaut Wendy Lawrence told me that she sleeps well in space as long as she isn't cold. "I would much rather sleep in space," she says. "I get a much better nights' sleep in space than I do down on Earth. You can just go inside of that sleeping bag, relax, and it's incredibly comfortable."

Presumably, if you were a space tourist sleeping with your loved one, your hotel would provide you with a sleeping bag big enough for the two of you to sleep comfortably. You would just have to make sure that neither of you floated out of the entrance of the bag. It's easy for one person in a sleeping bag to draw the top closure smaller than his or her shoulders. It is harder to do with two people in the same bag. Either they would have to remain with their heads quite close together (cozy – but a problem when you inevitably bump heads) or be tethered to the bag some other way. Of course, you can always sleep in separate bags, and snuggle in the same bag together when you are awake, which wouldn't be quite as nice. Don't you agree?

A better solution would be a double-headed sleeping bag. Imagine a sleeping bag large enough for two people. Now add two stretchy t-shirt collar-like holes about a foot apart to the top of the sleeping bag. You would get in and out through the side of the bag that would zip shut from the inside. That way, couples could play together in the sleeping bag while keeping any bodily fluids inside, and then fall asleep together too. The bag would have holes for ventilation, and might even need some sort of quiet fan sewn in the bottom to increase ventilation during *hot* encounters.

Getting Some - Peace and Quiet, That Is

Now and for the foreseeable future, spacecraft, whether a Shuttle, station, or hotel, are not quiet places. The hum of machinery, fans and other equipment necessary for survival are constantly a bother. The noise is comparable to driving down the freeway at 60 miles per hour with the windows rolled down! As you probably

expect, many astronauts today use earplugs to drown out some of the constant noise that bombards their ears. You might want to take a pair with you on your space voyage too.

Many astronauts don't sleep well for the first few nights in space. This may be due to motion sickness, the unnatural lack of a bed pressing against one's body, the excitement of being in space, or simply the fun of spending hours looking out the window at the fantastic view of the Earth and stars. Some astro-

STS-26 Astronaut Dick Covey, sleeps wearing eyeshades and a headset, propping his feet under the pilot's seat with his head on the on-orbit station panels

nauts have to take sleeping pills just to get some rest. It may be a good idea for space tourists to pack some, as long as their doctor agrees.

Presumably your space hotel suite would have a window to view the stars and the Earth below. The problem is that the Sun "rises" every 90 minutes when you are in any sort of craft in Earth orbit. Blackout window shades, or eyeshades like many astronauts wear, would be advisable.

Many astronauts have described a profound sense of spiritual and philosophical awareness from having experienced the space environment and taking in the incredible views of the Earth and space. Some have noted the absence of political boundaries when viewing our home planet from such a unique vantage point, and many have had keen interests in Earth's environment raised by their experience. It seems clear that being in space will awaken, or re-energize, a person's spiritual side. For many this feeling of peace and connectedness with the Universe may increase the libido.

In an article that appeared in *Fantasy and Science Fiction Magazine*, Sci-Fi Novelist Ben Bova wrote, "Imagine two human bodies floating in zero-gee. The slightest touch on your partner will send him or her bobbing away from you," he explained. "You are going to need a compartment with padded walls, or you will have to confine your lovemaking to an intimately small cubical where there is no room to roam."

Love in Space. Courtesy Space Island Group

Your future space hotel suite will be well designed for romance and play. I'm sure that you agree that if you can afford a place in space, then it should be well suited for weightless sex. Guests like you will demand the best experience money can buy. We all know that every couple that travels into space is going to try sex in space, so hoteliers are going to be sure and give guests the best environment in which to do what they want. You will likely see nicely decorated padded walls, a big porthole to the outside (one way for privacy, of course), and lots of straps, footholds, and handholds to grip and hold. No doubt there will be a waiting list for hotel suites as soon as word gets out of the wonders of very personal spaceflight.

This room would be yours to play in with your partner. Spinning and tumbling, or just floating with each other. Explore weightlessness as you float chocolates across the room to each other, and then sip wine from a plastic bag with a straw.

Space Activist Derek Shannon pointed out that the room would have to be designed to deal with fluids, "While all sex is messy, this could be disastrously true for zero-g."

"Facilitating sex in zero-g will require some creativity," says Space Activist and Business Researcher Amanda Harris. "Companies could develop special suits or devices that could assist a couple having sex in zero-g." In Chapter 5: *To Infinity and Beyond*, we'll imagine the ideal space hotel room for intimate encounters. First, let's consider what might be required to achieve various sexual positions in space, and of course, what to wear.

What to Wear

One must have the proper outfit for both launch and landing. Those bright orange suits that astronauts wear for launch, reentry into Earth's atmosphere, and landing, are also useful in case of emergency. These unique suits are made of flame retardant material, and are sealed to provide air in case of pressure loss inside the spacecraft. Astronauts wear these suits for several hours before, during, and after launch. Harvey Wichman suggested that removing these suits soon after achieving orbit, might actually be part of the cause of space sickness (SAS). Reportedly, when the astronauts on the Shuttle take off their launch suits, the whole cabin wreaks of body odor and ammonia. "If you're going to get sick that's when you do it," he says.

Space tourists will be trained to use similar suits, and diapers, before leaving the Earth for the celestial pleasures. It's a necessary safety precaution. Just like the astronauts, once in orbit, vacationers will be able to wear "shirtsleeve" clothing like lightweight pants and shirts. But for tourists, the romance of space will create the desire to wear something sexy.

The important thing to remember is that in weightlessness clothing will not dangle downward. So skirts, robes, capes, and similar hanging clothes will float everywhere – up and down. This may be embarrassing in public places, so you would want to avoid wearing them in the hotel lobby. But, in private it may be fun to have your skirt float "up" and "down" depending on your movement and the breeze. It would be fun to put on a cape to "fly" like Superman as you launch yourself from wall to wall. It's not all about sex, you know.

Since your space hotel suite would probably be warm enough, you may not need to wear anything at all. Think of the possibilities. You could float naked in the middle of the room looking out at the Earth and stars. Or wear super sexy skintight clothing, something like a wetsuit may be the fashion; fine to wear in-room or in public areas. But only those with a great figure would be willing to wear it. Vanna Bonta thought the opposite would be more enticing, "Wearing billowy clothes that float, and dancing for one's lover would be romantic."

Fashion Designer Janet T. Planet sees distinct groups of people who will be interested in space wear. The first space tourists will be trend setters and thrill seekers. She told me, "For "New Trendies/Thrill Seekers", who are used to wearing designer clothes typically, suits that flatter and have references to designer and performance clothing will resonate." She explains, "This "designer" group will look for comfort and ease of movement, referencing sports gear, with sleek clean lines. I would choose many performance fabrics, but use interesting yet understated cuts to maintain a futuristic minimalist aesthetic."

As space tourism becomes more affordable and more people travel into space for pleasure, another group will emerge. One that Janet T. Planet calls "Fiction Realists." She says, "I expect that this group will look to realize their dreams of futuristic sci-fi space adventures with space tourism, with their clothes playing a major role in this.

For designers," Planet explains, "there is endless inspiration to be taken from years of brilliant costume description in literature, and costume design in film and television, as well as from the work of futurists. For the Fictional Realists, asymmetrical cuts, sculptural forms, metallic fabrics and flashes of bright color will be cues of futuristic sci-fi style. These are all elements that I traditionally include in my designs, but having the backdrop of space will somehow make them more permissible for people who enjoy wearing them. It is as if space is a blank canvas with few preconceived style references and conventions, so wearing what you want to wear, even if it involves a fantastical futuristic suit, will suddenly be deemed as more socially acceptable," she says.

As this newfound freedom to wear more unusual clothing in space takes place, it will be reflected in Earth fashions. In the future, space clothes will set the trend for fashion here on Earth. I still like the backless orange tank top that Bruce Willis wore in the sci-fi adventure, *The Fifth Element*. Not many men can wear that and still look macho.

A scene from George Griffith's 1900 novel, *A Honeymoon in Space*, where women and men wore fussy Victorian clothing. None of that for future space tourists.

Getting married in space and don't know what to wear? No problem, Japanese fashion designer Eri Matsui designed a wedding dress, "that looks good without the aid of gravity." Matsui told the British newspaper *The Guardian* that she photographed models on parabolic flights sponsored by the Japanese space agency JAXA to help her design for zero-g romance.

There should be plenty more spacey outfits for the fashion conscious to choose from in the near future. Eri Matsui is also running a "Hyper Space Couture Design Contest." The winner will collaborate with her on designing clothes for the "first generation of fashion-conscious space tourists." Rocketplane Ltd., a launch vehicle company, is a sponsor of the competition.

Of course, clothing made specifically for lovemaking in zero-g will be all the rage. If you did a strip tease in weightlessness, your clothes would float around with you. Adding to the sexy atmosphere, your empty clothes wouldn't drop to the floor, but would take on a life of their own and float around you. It might be a distraction to some, or it might enhance the surreal atmosphere to an extraordinary experience. Later in this chapter, we will talk about those items made specifically to keep a couple together when having sex.

Docking Maneuvers: Sex That's Out Of This World

This section and the next few are more explicit about the act of sex itself. Feel free to skip it if you are easily offended. But then, if you are easily offended, then why are you reading this deep into my *Sex in Space* book? Sex is an important topic but it also can be an embarrassing one. Stay with me and we'll try to avoid the giggle factor a bit. Not much, but a bit.

I'll be blunt. When a couple has sex here on Earth, gravity plays more of a role than most people realize or admit. According to physics, and some guy named

Isaac Newton, action causes reaction. So when you thrust your hips towards someone or something, the rest of your body moves away such that the center of gravity stays the same. This is true *unless* something prevents the rest of your body from moving, such as friction from a bed, your partner, or some sort of restraining device. On Earth, you use gravity to bring about friction between yourself and the bed, or any surface that you have your weight on. Then you can push against the bed or other surface to move yourself forward. In space, however, that's not possible unless you are holding on to something.

Momentum is conserved, so says that Newton guy anyway, and any physics experiment will prove this fact. Momentum is the tendency of a body in motion to stay in motion, and for a body at rest to stay at rest. Even though you are weightless in space, you still have mass, and therefore still have momentum. So even though you can, for example, pick up a car and throw it across the room (assuming you are braced against a wall – remember the action of throwing a car will result in pushing you back), you can't throw the car fast, or stop it quickly if it is going fast, since a car still has lots of momentum due to its large mass.

So when you thrust against your partner, he or she is going to have a tendency to bounce away. If your partner is against something, such as a bed, floor, wall, window, whatever, then that holds them in place. On Earth, if your partner is on top then gravity will hold them down and onto you. However in space, your partner would either need to be up against something, restrained, or holding on to you in order to not be pushed away when you push towards them.

So in general, how do the laws of physics change sex as we know it? Well most significantly, a couple would have to hold on to each other tightly or lose their "docking." But it shouldn't take much to hold on to each other. In a 1973 *Sexology Magazine* article titled "Sex in a Spaceship" Isaac Asimov described it this way, "Under zero-gravity, the grip of arms and legs will be quite sufficient to assure all the contact desired. In fact, under zero-gravity, contact would be improved, for then intercourse can take place in mid-air and there is no necessity to be aware of the feeling or pressure of anything but your partner," he wrote.

Space Architect and President of the Space Tourism Society John Spencer agrees, "I don't think it's going to be very hard to have sex in space as long as one person is hanging on to something – the other person – or whatever," he said.

"I have no doubt that people will figure out a way," says Harvey Wichman. But he believes it will be more difficult than just grabbing on to your partner. "I have talked to scuba divers who have done it in the water," he explains. "Their descriptions are always 'well; the woman got one knee around a ladder, and hooked an ankle around another ladder, and held on with two hands and then the man held …' You see what I'm saying it is possible to do. It wasn't floating about and doing it. It doesn't work like that. You're going to have to get yourself mechanically constrained somehow," he says.

Most people that I spoke with for *Sex in Space*, however, didn't see any physical reason why lovers couldn't just float in the air and hold on to each other. But I suppose it's going to take some testing to prove one way or another. "I think it's a lot easier than most of us on the ground make out," said Space Activist Derek Shannon. "It is the initial awkwardness that is most likely to detract from the romance. So it will take practice to make perfect," he says. I agree completely.

No one is sure what sex in space would be like. If anyone has done it, they're not talking about it, at least with me. So opinions vary. Space experts James and Alcestis Oberg wrote that astronauts who try sex in space, "May thrash around helplessly like beached flounders until they meet up with a wall they can smash into." But science fiction writer Ben Bova is more optimistic. He wrote, "If two orbiting spacecraft can be mated by remote control, a human couple with grasping hands and willing minds should be able to solve the problems of rendezvous and docking gladly. Solving this problem should, in fact, be quite enjoyable."

Aerospace & Human Factors Engineer Juniper Jairala, has very long hair and pointed out another advantage of zero-g. "It would be great to not have my hair in the way. To be on top, and have my hair floating away from my face, would be great, as opposed to hanging down." She explains, "I recall being on the Zero-g aircraft and I had my hair loose, and I don't do that very often because it drives me nuts; it's heavy and gets all over my face. But in zero-g it never bothered me once. It was always just away from my face. I didn't even know my hair was there," Jairala said.

Cosmic Kama Sutra

What sexual positions will work in space? There are some Kama Sutra positions that, from the pictures, look like they would require you to be a yoga contortionist. These positions would still be just as hard in space, and most of them wouldn't work without something to hold you down or in position. But there are other positions that a lack of gravity would make easier. For example, if each person is facing a different direction, that would be easier in space. In his 1997 thesis on sex and spaceflight, Sexologist Dr. Ray Noonan joked, "I once heard that several of the positions in the ancient sex manual of India, *Vatsyayana's Kama Sutra*, can only be done in space!"

Former NASA Consultant G. Harry Stine in his 1985 book, *Handbook for Space Colonists* noted that, "There isn't a single one of the legendary "thousand-and-one ways of the ancients" that won't work in the weightlessness of space." Well, I'm not convinced that's true, at least without a lot of bracing. But I am sure that at least several positions are possible where the couple is holding on tightly to one another. And even some positions where someone is "upside down" would definitely be easier.

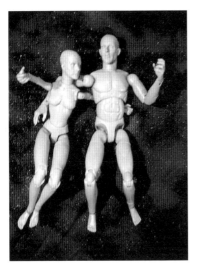

Barbarella and Buck float and
enjoy space together

Let me introduce my two manikins, Barbarella and Buck. They're not anatomically correct, so we'll just have to pretend that they are. In this first picture, they are drifting side by side looking at the stars through their space hotel suite windows. Now, Barbarella and Buck will show you a few positions that should be possible in the weightlessness of space. This is not intended as an exhaustive list, but will give you some idea of what should be possible. Most of the positions are for a heterosexual couple, but some would be useful for homosexual couples too.

We see Buck and Barbarella drifting towards each other from opposite sides of the cabin. They grasp hands and try to cancel their momentum without inducing a spin. It's tricky and will probably take them several tries to get it right.

Next we see Barbarella and Buck embracing for a kiss, after all some foreplay is crucial in all lovemaking. They are hugging with one arm only here, so we can see how their legs interlock so that the torsos stay together. Hugging with just arms may not be enough for a full body contact.

Lovers float towards each other grabbing
hands to cancel one another's momentum

Hugging in space. Notice the
legs also lock to keep the
couple together

Soon things with Barbarella and Buck have become pretty hot and heavy. Because of air currents, the couple has probably drifted towards a wall (unless the room is huge) and just lightly bounced right off. They bounce every so often now and, for obvious reasons are almost oblivious to their surroundings. They've chosen to keep their first encounter simple with the basic position, a variation of the 'missionary' position on Earth. Barbarella's legs are wrapped around Buck's buttocks and they are holding each other with their arms and moving as desired.

Basic sex in space position, a modified missionary position

If Barbarella and Buck weren't holding tight and her legs were not around him, then every pelvic thrust from either partner would cause the other to fly away and "disengage." That would undoubtedly ruin the mood. If they both had their backs to a solid object, (like a tube – see *Sex Toys* below) then that would keep them in place.

Holding on tightly to the upper body is probably not enough, so it seems that holding tightly with the lower body is probably necessary too. By this, I mean that Barbarella has her legs wrapped around Buck's torso. I say "she," because I think that if a man tries to wrap his legs around a woman when in this pose, it's probably going to be impossible for him to keep his manhood in position. However, just holding on with the lower body isn't enough either. With each thrust, Buck is going to have to pull Barbarella down onto him, probably by holding her by the shoulders; otherwise Barbarella will just float away.

You can see why free floating sex might be a bit tricky. Our couple has to modify the standard face-to-face position they might use on an Earth bed to make it work in space. Momentary slips do not cause immediate "disengagement" as long as there is "redundant" gripping.

As a variant of this position, lets say Barbarella has her back against a (preferably padded) solid surface such as a wall, ceiling, floor, window, structural beam, etc. Then Buck could choose to hold on to some sort of handgrips and/or foot grips on the surface behind Barbarella, instead of holding on to her. Essentially Buck would pin Barbarella to the surface behind her. Or, Buck and Barbarella could switch positions. Space is an equal opportunity place, you know.

Immediately we see the advantage both of padding (to prevent bruises when bumping the wall) and cloth straps at many conveniently located places on the wall

Weightless 'sitting' position with legs locked

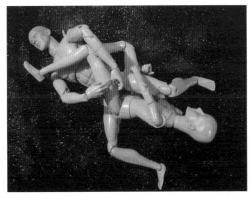

Sexual position with Interlocking Y legs

Sexual position facing opposite directions

to grab onto. Free floating in space is a bit risky when "engaged" because air currents and other things could knock our happy couple around. All furniture and objects should be removed and everything should be padded for your protection.

On another encounter, Barbarella and Buck decide to have sex in a sitting position. This can also be done free floating, as shown in the photo. Notice that Buck and Barbarella are using their arms to hold onto each other and her legs are wrapped around his.

On the next encounter Buck and Barbarella decide to get a little more creative, just to see if something that may be a bit difficult on Earth is possible in space. I'm not sure if this position would work for everyone, because the man's member is forced downward while erect. However it does show that the two people don't need to be facing in the same direction. In this position, Buck and Barbarella's legs make two interlocking Y's. Each of them has a firm grip on the other's leg, and if their timing is right, they are pulling together rhythmically. This forces them together with each pull, and they spring apart with each relaxation.

When Buck and Barbarella have sex next time it's similar to the missionary position, but swiveled 180 degrees using the

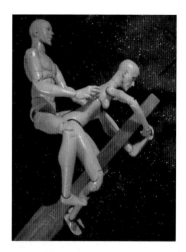

Riding the rocket, showing a vigorous spooning position

Missionary position with an elastic waistband holding hips together

penis as the axis of rotation. In this position, they each grab the legs of the other person. It's difficult for Buck to achieve a good penetration as Barbarella's legs are too close together. Each has to thrust their pelvis towards the other at the same time or else they will disengage. This doesn't seem to be a very successful position, but it does show the variation in orientation. Not every experiment with sex in space is going to work. But as I have learned from *The Magic Schoolbus*, we learn by taking chances, making mistakes, and getting messy.

Next time, Barbarella and Buck decide to try something they are sure will work, let's call it "vigorous spooning." Buck approaches from behind while Barbarella holds onto something, it could be the wall, straps for hands and feet, whatever. In this case it's a structural beam. It doesn't seem that free floating is an option in this position. Barbarella must hold on to something with both hands and feet for this to work. Buck only needs to hold on to Barbarella and do his thing.

There are various suggested sex toys that can be used to hold Barbarella and Buck together during zero-g sex. In this next example, the couple uses a simple thick elastic band to hold their hips together. This allows Buck and Barbarella to move their legs freely without disengaging. However, they still must still hold on to one another's upper torso and pull down countering every upward thrust.

This next position may seem a bit like bondage, but will help Buck and Barbarella enjoy each other's company in weightlessness. The space version of straps is tethers, and they are quite useful in space. As Harvey Wichman explained, "Restraining devices have been developed for physicians who have to examine somebody in space…something like that can be used [for sex]," he explains. "There are social implications for strapping people down when having sex, that some people would be uncomfortable with. On the other hand, maybe they wouldn't be so uncomfortable because this would be an excuse for it."

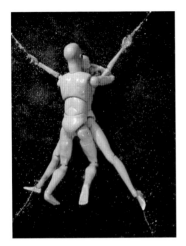

So here we see Barbarella tethered to the wall as Buck gets it on with her. It is his responsibility to keep the two together. However, there isn't the risk of her floating away and disengaging. Also notice that they are not bumping into random walls anymore.

Buck mounts a
tethered Barbarella

Of course, what is good for one is good for the other too. In this picture, Buck is tethered to the wall as Barbarella has her way with him. Now it's her responsibility to stay together. A gay or lesbian couple might find an advantage in using bondage too and would position themselves as desired.

Barbarella mounts a
tethered Buck

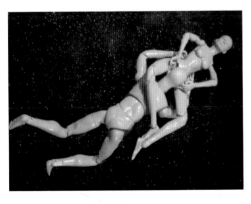

Buck and Barbarella demonstrate oral
sex in space. Notice that they have
to hold on to each other

Next Barbarella and Buck try oral sex. They can do this in freefall, but the key is for the person giving oral sex to hold the hips of the person receiving the pleasure. Oral sex in space might be easier then here on Earth. One could position their partner in such a way that access and satisfaction may be greater. We assume that Buck and Barbarella will switch off pleasuring each other this way.

The classic "69" position in space

The basic position with help
from the third dolphin

As expected, they repeat the oral pleasure with the classic 69 position, each making sure to hold on to the other. As Derek Shannon put it, "I think sixty-nining in particular would be much easier, with the biggest worry is drifting off somewhere inappropriate."

One speculation has been that a third person would be very useful in keeping the other two together. This is the "three-dolphin" situation made famous by G. Harry Stine that we talked about in Chapter 1: *Has Anyone 'Done It' in Space*. Here we see Buck and Barbarella doing it again, this time a friendly dolphin is helping to push them together. No doubt the dolphin hopes to get his jollies later. While this may be to some people's tastes, many will not be open to having a third person around, much less a dolphin. But like the use of tethers, in some cases, perhaps it's just the excuse the couple was looking for to invite the dashing or beautiful captain to join them.

In general, any position that involves the couple holding on to each other as tightly as possible will work. And any position in which they are pushing apart, that typically relies on gravity and friction, may not be possible without restraints of some sort. There are many more positions are possible, and part of the fun is speculating what will, and will not, work. We won't know for sure until we get some brave volunteers into space, do some serious sexual experimentation, and report back to us with the results.

Toys in Spaaaaace

Most of the sex toys available on Earth would work just as well in the weightlessness of space. However, don't let them get away from you. *Wired* Sex Technology Columnist Regina Lynn, had a method for keeping them handy, "Any toy you can use in gravity you can probably use in space as long as you hold on to it," she told me. "Maybe add a leash, like boogie boarders do, so if you let go it doesn't fly off."

There are a few unique items that would serve little purpose on Earth, but that lovers might find useful in space. These are generally items that help to keep companions together in weightlessness.

Tethering a partner to straps, the space equivalent to tying your partner to the bedposts, is seen by some as rather kinky, but it would actually help quite a bit in space, as shown by Buck and Barbarella in the previous section. The partner tethered in place would not have to worry about floating away when being bumped, or rather banged, by their companion.

Lovers might want to tether to each other to keep close. This would allow a quick reengagement if there were an accidental disengagement.

You could have sex in the sack too. Many co-ed campers will tell you it is possible to have sex in a sleeping bag, as long as the bag is big enough. The two-headed sleeping bag mentioned earlier in this chapter is perfect for a morning rendezvous. Since the sleeping bag is tethered to the wall to keep it in place, then there isn't the concern about bumping walls or getting too far from your partner. If the couple disengaged accidentally at any point, the sleeping bag would at least keep them from floating too far apart so they could get right back to the action.

A tighter version of the sleeping bag specifically designed for sex in space, was suggested by Space Tourism Society Vice President Samuel Coniglio. "An item I call the Snuggle Tunnel," he explains, "is a four-foot diameter tunnel with lots of padding and fur. The tunnel ends in a view port so the couple can enjoy the Universe and each other's company. The tunnel gives the couple room to play yet they don't bounce away too far," he says.

This space tube could also be inflatable so that it would be easier to pack and carry to an orbital hotel. A similar device was recommended in the phony NASA report about sex experiments in space that was mentioned in Chapter 1 and is included in the appendix of this book.

Harvey Wichman thought these devices were ridiculous, "The sex sleeve that holds two people together in a sleeping bag kind of thing, they'd croak in that. That's not going to work." He explained that when people have sex, they put out a lot of body heat. On Earth that heat rises, even through blankets and sheets. But in space heat doesn't move unless forced by air currents. Therefore, Wichman believes that the heat buildup in a 'sex sleeve' would be too much for the couple if air couldn't circulate through. One solution might be to add a fan, or multiple small fans, to the sleeve and make sure the material breaths. I'm sure orbital hotel designers could get on top of this design issue.

Many have suggested items of clothing that would keep the couple together. One idea that I came across consists of a unique pair of "underwear" with four legs; like two pairs of elastic tight bicycle shorts put together strategically. Of course, this would not allow a couple to try other positions or move around much while wearing them. But it would keep lovers from accidentally disengaging. Who would make them? Of course, space lovers could have them custom made, but I can only imagine what being measured for a fitting would be like.

"No space-wear wardrobe should be without my invention of the 2-Suit," says Vanna Bonta. "It's the Love squared of fashion, and kind of poetic really because one suit envelopes two who will, after playing in it a bit, merge as one."

"When it's time for lovers to slip into something more comfortable, they slip into the same suit," Bonta explains. "It comes in fabric texture of preference: silky, latex, metallic chain link. Lovers can slip into it nude, or with layers to unwrap within. Their arms and legs share sleeves. There's ample room to move freely within if desired, and the inside is lined with belts and harnesses that can tighten or lessen the proximity of various points of the torsos and bodies to one another, or not. The 2-Suit can also fasten to a stable wall, from arms, legs, backs, or not. Roominess of the 2-Suit is adjustable from within, and can be made smaller via a series of velcroed folds and pleats. It also has quick disrobe function, leaving only the harnesses in place," says Bonta.

As Buck and Barbarella demonstrated, however, all you really need is a wide strip of elastic sewn into a cylinder; like a giant scrunchie. Oh wait, they were wearing a scrunchie, but they're only a foot tall.

You might find Velcro-like items of clothing useful on your voyage into orbit. Velcro-like slippers could be used to keep a grip on the floor. Or a shirt with a Velcro-lie back might keep one person attached to the wall for a more traditional Earth sex position.

Maybe a belt with handles or grips would be useful to grab onto your partner. But the belt would have to be tight. Or perhaps a belt with suspenders would be more secure without having to be so tight. Very quickly many of these suggestions start to look like an advertisement for sexual bondage hardware, because holding on to your loved one in space requires strapping them to you.

Derek Shannon thinks that most of the space sex toy ideas are too much, "I don't think there will be any elaborate gear needed beyond what enterprising astronauts must already improvise on a regular basis. But since the average space tourist will be somewhat less resourceful, I'm thinking that the first astronaut to break his or her silence [about having space sex] will have a lucrative branding opportunity for a line of effective, but still simple, aids," he says.

Of course, if you bring lubricant, it should be a thick gel or cream, not a watery oil. This is because liquids don't coat surfaces well in space. Surface tension causes liquids to bead up rather than soak a material or skin.

A condom should work just as well in space as on Earth, but couples will need to be very careful when removing it because the spillage may float away, and will not be as easy to clean up as it is on Earth. The condom would need to be removed before, ahem, shrinkage occurs. If not then the liquid may come out more easily than on Earth because gravity wouldn't be around to hold in the liquid inside the condom.

As the technology for three-dimensional printers is perfected, space hotel guests could create items and toys on-orbit on demand. A three-dimensional printer selectively binds plastic dust together in a controlled box to build a solid object up layer by layer until the object is complete (yes, early versions of this really exist). Who knows what creative items could be ordered, used, and recycled as needed? Of course, the instructions for the item would have to be in a special digital format. But beaming information to space is cheap, much cheaper than hauling items to space. This exciting possibility would reduce the need to carry certain items into space. It might be used to make custom-fit, disposable clothing for guests each morning. In space, washing clothes is impossible, so replacing clothes is the way to go. Clothing, bedding, and even sexual toys and aids for zero-g could be created and destroyed as needed with no mess or fuss. Imagine the possibilities!

It would be useful to have your toys glow in the dark, especially if not tethered to you. This would allow you to see them after the sunset, which occurs every 90 minutes when the orbital vehicle passes behind the Earth blocking the sun for a little while.

Not having to ask anyone other than the computerized butler, for a sex toy or aid keeps things private, which appeals to most people. 'Computer, one double-headed sleeping bag lined in electric blue faux fur please.'

So, as you can see, weightless sex really is space tourism's "killer app." Now on to one possible result of lovemaking in space: Making space babies!

Chapter 3: Making Space Babies: Conception, Pregnancy, and Birth

Now that we've talked about the nitty gritty of sex in space, we've got to talk about what comes next. Babies!

This is a more serious topic, and is not such a light one, because everything we know so far indicates that it is probably quite *unsafe* to be pregnant while in space. As a mom myself, I'm very concerned that once people start having weddings, honeymoons, and vacations in space, that a zero-g lovemaking session will unintentionally, or intentionally, lead to pregnancy. It would be very upsetting to me if the result were a malformed child, or a child who could not be brought to term. The possibility is very painful to me. So until there has been more research and we have determined what is safe and what is unsafe, we need to take precautions. I believe that until more is known about pregnancy beyond Earth's surface, we should strongly discourage anyone from getting pregnant in space.

Once space tourism really gets going with frequent tourist flights, there is going to be a lot more sex in space than there has been, if there has been any so far. It is important to answer critical questions about conception before we have to deal with the implications.

Does birth control work? Can someone get pregnant in space? What are the risks? What does radiation do to a pregnancy? What does the lack of gravity do to pregnancy? If it is determined that a pregnant woman should be sent home to Earth, then at what point in the pregnancy must she return to Earth? If the mother can't get back to Earth, what can be done to protect her and the baby? What would a delivery be like in space? Would a Cesarean section be possible? And what about people who are in space for longer periods, can they have a normal Earth pregnancy and birth later?

Unfortunately, scientists know little about the effect that space and weightlessness have on the human reproductive system. In this chapter we will talk about what we do know, or can guess, about the possibilities of reproduction in space. Once this is certain, if people really are going to work and live in space someday, then we need to learn more. NASA Flight Surgeon and Psychiatrist Dr. Patricia Santy once wrote, "Can a child develop normally in zero-g. How is a woman going to deliver it in that environment? We need to know a lot more."

NASA and the Russian Space Agency have both done a number of animal experiments on reproduction. However, no human studies have been conducted. We can learn a lot from these animal experiments and based on them, we can make good guesses about human reproduction in space. Some of the results of this research with animals is disturbing and suggests there might be significant risks to the fetus if a woman should get pregnant while in space.

"[I'm] surprised spaceflight is legal," said Lynn Wiley back in 1989. At the time she was an Associate Professor of Obstetrics and Gynecology at UC Davis. "If you tried to get people to work in fields or in office buildings with as little knowledge of what working there is doing to their health as we have of what spaceflight is doing to astronauts, you'd never get away with it," she said.

Even with new biological experiments on Mir and the International Space Station, things have improved only slightly. Unfortunately, very little of that biological research has been on reproduction.

One of the things we have learned is that weightlessness has unexpected consequences on the body, beyond what is immediately visible. For example lymphocytes, our disease fighting white blood cells, are seriously damaged by weightlessness. Only three percent of lymphocytes showed normal activity on one Skylab experiment! So we know that once in space the body's immune system is potentially crippled and travelers are probably much more susceptible to disease than at home on Earth. This was certainly true when we first began flying into space. The mechanism for this massive reduction in white blood cells is being studied, but is still not determined or reversible.

"Seriously speaking, we know little about the role of gravity in fertilization and fetal development," former Astronaut and Director of the Laboratory of Cell Growth, Dr. Millie Hughes-Fulford told me. "We do know that certain body systems, bone and immune system require gravity for normal responses. In fact, we have recently finished work where we found that certain signaling systems, at the cellular level, do not function properly in zero gravity. So I would say that it would be dangerous to conceive and carry out pregnancy without normal Earth gravity," she cautions.

The Female Side

Millions of Earth women take birth control pills. One of the key questions for women in space is; will they continue to work? According to an article in the 2003 Annual Review of Medicine, many drugs act differently in weightless conditions. The full effects of various kinds of birth control pills have not been examined for effectiveness in zero gravity, even in animals.

Many women astronauts have been using birth control in order to avoid having a period while on a spaceflight. Access to personal medical records isn't available, but at least anecdotally, the birth control medicine appears to be doing its job while in space.

Former Astronaut Susan Helms told me, "I just worked around [my menstrual cycle] and took medication to stop my periods completely." For Helms, at least, the medication, which was probably in the form of a standard birth control pill, seems to have worked, at least well enough to suppress her periods. This leads me to believe that birth control drugs will work just as well in space, but I would like to see studies done with different specific types and dosages before taking the chance myself or suggesting it to any woman. That study is not likely to happen any time soon, at least not with NASA funding.

Menstruation is more difficult for some women than others. Many women have fought to overcome the "stigma" of menstruation. Since 1960, women have had access to birth control medications that have the effect of lessening, regulating, or eliminating menstrual flow as well as the associated pain and discomfort. Women are now able to take charge of their menstrual cycles and have them, or not, as they wish. Modern medications seem to have liberated women from the "burden" of menstruation (for those who see it as a burden). There is the possibility, and some evidence, that the space environment will actually change the menstrual cycle itself.

"Earth studies of women confined to artificially lighted caves show their menstrual cycle stopping," said Dr. Richard Jennings in a December 1991 Ad Astra article. "It will be interesting to see if daylight every 45 minutes and weird artificial lighting is indeed a problem."

The weightlessness of space really doesn't make a difference when it comes to having your period. Some women astronauts have had their period in space and used the same sanitary products that are available here on Earth. I'm told they work just as well in space as they do on the ground. That's because absorbent sanitary products rely on surface tension to soak up the liquid. Aerospace engineers use the same principle when putting metallic 'sponges' in the fuel tanks for spacecraft to make sure that the engines suck in liquid and not gas.

Due to privacy issues, an astronaut's personal medical data is not reported and is not public knowledge. But the anecdotal evidence from interviews that I made suggest it is not an issue. Astronaut Janice Voss told me that dealing with your period in space was, "very similar to on the ground. Blood doesn't flow down, so the first day was more of a surprise," said Voss. "And the urine filters in the toilet have to be changed more frequently." But otherwise, things work just like here on Earth.

Astronaut Ellen Baker wasn't interested in controlling her cycle either. "You just deal with it on orbit like you would deal with it on any trip," explains Baker. "It's never interfered with anything I've done. I've never let it interfere with anything I've done, nor come in to any decision making process that I have. So, I just decided I've lived with it all of these years, I can live with it in space."

NASA's concerns about women menstruating in space came late in the game, as late as 1982. It had been four years since the first U.S. women astronaut candidates were chosen, and a year before Sally Ride's first flight. Medical consultants met with NASA doctors to discuss the matter and came to the conclusion that the answer to the "problem" of menstrual flow was drugs, a.k.a. oral contraceptives that can delay or regulate a woman's period, or reduce menstrual flow.

To NASA's credit, the space agency does not require, and has never required, any of its women astronauts to "fix" her menstrual cycle. A female astronaut is given the option to delay or suppress her menstrual cycle by taking medication. Some do and some don't.

Radiation is a different concern for women than men. The male testes are more exposed to radiation then the female ovaries, just because they are not inside the body. There is also some evidence that ovaries might be able to take more radiation than testicles. However, the long-term effects of exposure are more severe. Dr. Harvey Wichman, an expert in Aerospace Medicine told me, "When a man goes up and gets irradiated and has irradiated sperm, he can come back down and in a week or so with a couple of masturbations, get rid of all of his sperm and generate new. But a woman who goes up has all of the ova at birth and they are all exposed, and she's in a different situation than he is," he explained.

Sheryl Bishop, a Space Psychologist from the University of Texas had an easy solution for long term missions where the exposure is higher, "What is being discussed on the medical side is that anybody who is going for a long duration mission, should be past their intent to have kids, or they should bank sperm and eggs prior to leaving," she explained. "They should not be using their sperm and eggs on return."

The Male Side

Since libido in men is strongly influenced by the sex hormone testosterone, it is quite possible that men might not be as interested in sex as they are when on Earth. A European Space Agency (ESA) study published in 1998 showed that spaceflight causes a reduction in levels of testosterone in male astronauts. Testosterone levels in men were measured during, and just after the mission, and were significantly lower than just a few days before going into space.

"Expect male hormone levels to sag," said Western Washington Physicist Jim Steward in an article for the Chicago Sun Times. "Maybe it's the sunless days, or the food or being [away] from home," he said.

Stress is also a cause of testosterone reduction, so it's just as possible that this reduction in testosterone is caused by the stress of spaceflight. So the solution might be as easy (or technically as hard, from an engineering point of view) as making space travel less risky, and no more stressful than an airline flight. We will talk more about the psychological and social effects of space in the Chapter 4, Sex on the Brain and Lust in Space.

One of the well documented side effects of weightlessness is that men seem much more likely to wake up with an erection. This happens because the body has excess fluid, especially early in flight. Former Astronaut Mike Mullane, in his book Riding Rockets explained, "I had an erection so intense it was painful. I could have drilled through kryptonite. I would ultimately count fifteen space wake-ups in my three Shuttle missions, and on most of these and many times during sleep missions my wooden puppet friend would be there to greet me," he wrote. We will discuss this again in the next chapter, but clearly, the good news is that men are physically capable of achieving an erection in spaceflight, even if testosterone levels are lower.

Are sperm affected in flight? There have been a few cases in animals where sperm have fertilized ova, so that would indicate they are functional. However, according to the article, "Effects of Space Travel on the Human Reproductive System" published in the August 1989 issue of the Journal of the British Interplanetary Society, the tails of dog sperm curl up and shrivel in space, just like a dog curling up for a nap. If true then sperm effectiveness may be seriously reduced.

Sperm motility, or the ability of sperm to move around, is important in reproduction because the sperm needs to be mobile not only to reach the egg but to penetrate and fertilize. Experiments with bull sperm in test tubes and with live sea urchins show that in microgravity sperm motility is increased. According to Dr. Joseph Tash, Principal Investigator of the sea urchin sperm experiments, "Despite the difference in species, sperm and sperm movement are universal in the animal kingdom." He chose to use sea urchin because they can survive pre-launch delays better than other species that have to be treated more carefully.

So you would think that this increased sperm motility in bulls and urchin increases fertilization, but that's not yet clear. Sperm movement begins with a chemical process called phosphorylation, in which an enzyme changes the functioning of a protein within a cell. This sets off a chain reaction which leads the tail of the sperm to move. On Earth, tail movement is halted or modified when a second enzyme, a protein known as phosphatase kicks in. But in space, the second enzyme, phosphatase doesn't do its job. This results in faster tail movement.

Dr. Tash explains, "If enzyme processes are being altered by gravity, and they are, you can't even guess at the effect on fertilization until you've studied more than just sperm movement."

The opposite is also true. Hypergravity, or the increased gravity due to a launch or other acceleration, decreases sperm motility. Based on further centrifuge testing by Dr. Tash, with hypergravity as low as 1.3 Earth G's, there is a 50 percent decrease in the rate of binding sperm to eggs, and also in subsequent fertilization. Since increased gravity is possible with a centrifuge on Earth, then the decreased fertility was also confirmed with experiments on Earth. Unfortunately, the low gravity fertility experiments can only be done in space.

No one knows how gravity affects our cells. You might think that something that small and seemingly immune to orientation would be unaffected by gravity. But from the experiments on sperm enzymes, as well as experiments on statolith cells in plants, it is clear that gravity does affect these microscopic structures too.

Since there have been no onboard experiments, it's not clear what happens to human sperm in weightlessness. Instructions for an in-flight experiment on human sperm collection would, no doubt, be quite amusing. Containment of the sample would be critical. Imagine what the flight manifest might contain for visual aids requested by male crewmembers to help collect samples for the experiment.

No studies have been conducted to evaluate the effect of space on human spermatozoa, both while in space and the production after returning to Earth. However, based on the number of male astronauts who have become fathers after returning from space travel, it seems that the long-term effects, if any, are minimal.

Sperm cells in men are significantly more vulnerable to radiation than a female's egg cells, according to Dr. Richard Jennings, former Chief of Flight Medicine at NASA's Johnson Spaceflight Center in Houston, Texas. A rem is an arbitrary unit of radiation used to measure exposure to humans. Typical Shuttle missions lasting only a few weeks get 1 rem of exposure, which isn't excessive. This is in low Earth orbit, but it takes as little as 10 rems to begin to trigger a reduced sperm count, and 50 rems will result in temporary male sterility. Of course, one rem is typical on a 10 day Shuttle mission. Longer missions, or missions outside low Earth orbit have much higher radiation dosages.

Radiation

Radiation is an extremely important issue for astronauts and space tourists, especially when considering something as important as making babies. Let's look at what radiation is and how it affects our body.

Small atomic particles, such as protons, neutrons, and electrons (or even larger atomic nuclei without their electron shells) that are moving at high speed are all labeled "radiation." Photons, which are massless particles that behave like a wave and move at light speed, are another form or radiation.

Radiation causes damage by colliding with molecules and causing a reaction.

This becomes a problem when radiation collides with the DNA that makes up our bodies. Basically, DNA is the molecule responsible for genetic transmission from one generation of cells in our body to the next. DNA is a very large molecule and it can be ripped apart by collision with different types of radiation.

Some damage is not a problem; we all absorb some radiation all the time. But, with enough DNA damage, the body starts to have problems. In addition, radiation hits other chemical molecules, which cause them to shed electrons and become free radicals. Free radicals are chemicals which have a charge and want to latch on to other chemicals. The presence of free radicals in the body can upset the natural body chemistry, causing a chemical reaction that is not desired. Again, the body can handle some level of this unexpected chemistry, but too much can overwhelm the normal responses.

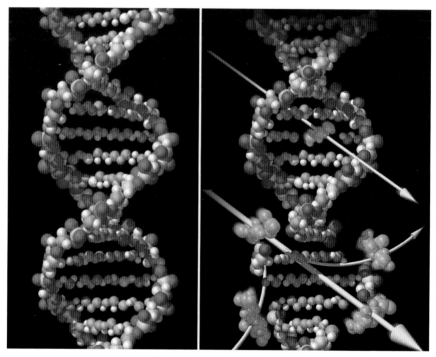

Normal DNA (left) and exposed to radiation (right). X-Ray damage at top right, and heavy ion particle damage at lower right. Illustration courtesy of NASA's Dr. Frank Cucinotta, NASA/JSC, and Prem Saganti, Lockheed Martin

There are several radiation sources in space; cosmic radiation, magnetosphere radiation, solar eruptions, and proton radiation from the Sun. Here is a short explanation of each, so we know what we're dealing with.

Cosmic Radiation consists of all different types of particles that are floating out there and permeate our Universe. Cosmic radiation is made up of protons, neutrons, and electrons, as well as the charged nuclei of many types of atoms. This type of radiation is constantly with you in space, although on Earth we are protected by the atmosphere and magnetic field of our planet. It's one of the nice benefits of planetary life.

Planets with a conductive molten core, such as Earth, Venus or any of the gas giants like Jupiter, have a large magnetic field. This field protects all life on the planet from radiation by trapping charged particles from space within the field. Earth's magnetic field is called the Van Allen belt. The belt is a region that presents a dangerous area for space travelers to travel through, as that is where many charged radioactive particles are trapped. While in that region of the magnetosphere, spacecraft and people are subject to a lot of radiation from charged particles, especially high energy electrons. So spacecraft usually pass through it as fast as possible so as to avoid disruption to their electronics. Only the astronauts who went to the Moon have had to pass through it, and they did so quickly.

Solar energetic particles such as protons and neutrons are emitted from the Sun. Usually the level is low and constant but a solar eruption can cause a huge spike in solar radiation. This type of radiation travels much slower than light, so we can see the event on the Sun and have some warning before it reaches us. In most cases we can take some sort of shelter, such as underground on the Moon or Mars, or get down to Earth if in orbit. But if we are in deep space, such as traveling between the Earth and Mars, then there is nowhere to hide. A solar eruption can also cause disturbances in the magnetosphere causing more radiation from that source to bombard the planet and objects in low Earth orbit.

Photons, which we see as light, are a form of radiation given off by the Sun, or other sources such as a light bulb. Photons in many wavelengths such as visible light are harmless, or relatively benign. Sunburn is caused by photons in the ultra-light spectrum. But at some wavelengths, such as microwaves and x-rays, photons can induce reactions and cause cell damage. That's why the microwave oven works, photons in the microwave frequency cause molecular excitement primarily in the water molecules of our food.

The Sun puts out lots of photons of all wavelengths, including radio, microwave, infrared, x-ray, visible, and gamma wave frequencies. The specific temperature of our Sun means that most of the photons are in the visible and infrared wavelength, which is not a problem. Few photons, percentage wise, are emitted in the x-ray spectrum, for example.

Certain materials can block photons, and is easy to figure out what works for most wavelengths. But some very small wavelengths, such as x-rays, are very hard to block and take lots of dense material. Gamma rays are so small that they are very difficult to block, but that means they are also difficult to interact with and are

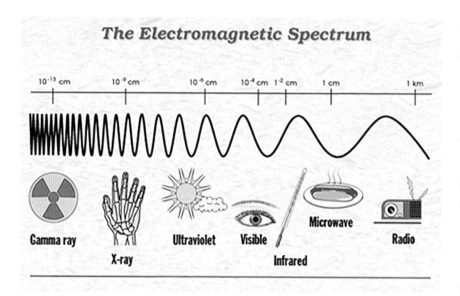

The electromagnetic spectrum. Image courtesy NASA

therefore considered to be essentially harmless except in extremely high dosages. X-rays are therefore the harder problem because they take a lot of material to block, but also can do some damage when they hit.

So, the more radiation (of all types) that a person is exposed to during their lifetime, the greater their risk of cancer and future problems as they get older. Radiation damage to cells appears to be cumulative throughout your life and the longer you statistically have "remaining" to live, the more chance you have that the damaged DNA can grow out of control causing cancer. It makes sense then that older people can take more radiation because statistically they are more likely to die of other causes before the radiation damaged cells can develop enough to cause a problem. The first Mars astronauts may be age 50 or more, just for this reason.

Now a young child's body which is still growing has a much higher risk of damage due to radiation. That's because every damaged cell and DNA pattern duplicates much more often when you are growing, and becomes a larger percentage of who you are. Or, the radiation can disrupt the reproduction of chromosomes in the developing body by making the chromosome breaks incomplete. This is called Mitosis. So radiation is particularly dangerous to a growing ovum, embryo, or fetus, because it is growing at a very fast rate.

Fortunately, the pregnant mother's body mass does help a lot in protecting her unborn child from most kinds of radiation. But some types, like x-rays, cannot be protected against as much by body mass. This is why doctors and dentists will not use x-rays on a pregnant woman unless there is a medical emergency.

Unfortunately, we don't know enough to say how much radiation exposure during pregnancy would cause a problem. In a 1991 Ad Astra article about reproduction in space, Dr. Richard Jennings said, "At the moment our knowledge indicates the risks are linear – the more radiation you get, the greater the risk of genetic damage, but we don't know where the cutoff lines are."

A rem is a unit by which we track radiation exposure. For a full term pregnancy the maximum recommended dosage is no more than 5 rem. If a fetus were carried from conception to delivery in Earth orbit, it would be exposed to as much as 30 rem. And that's while still protected by the Van Allen radiation belt. Outside of low Earth orbit, the dosage is much higher, and the implications may turn out to be terrifying.

Conception

To date, data on human reproduction and sexual physiology are limited due to an astronaut's medical confidentiality. So, let's take a look at the facts we do have and extrapolate.

It is interesting to note that the National Research Council recently stated, "Studies on the effects of space travel and the physiological adaptation to zero gravity have not yet addressed alterations in the reproductive system in either men or women." It seems amazing that a subject so important has not been studied.

Lynn Wiley a Developmental Biologist at the University of California at Davis once said, "Can men make good sperm in space and women make good eggs, and can that grow into a little baby? ... We really have no clue."

Sperm cells are guided to the egg via chemicals that the ovum emits. So the sperm doesn't need gravity to know where it's going. It seems there is no reason why a sperm shouldn't find the egg just as easily in space as it does here on Earth. And then theoretically the fertilized egg should develop, but it's quite possible that there is something we do not understand about the effects of weightlessness.

Harvey Wichman told me, "Nobody knows if the egg will embed in the wall [of the uterus], or if gravity plays a significant role in contacting and embedding in the wall. The one thing we do know is the Russians have not successfully managed to hatch normal quail eggs in space. This is what we should be doing on the dang space station right now - having animals, mammals, rats, and seeing what takes place. Otherwise we've got to make sure that [women] don't get pregnant when they are up there and end up having a deformed child or something," he says.

A fruit fly breeding experiment on a 1995 Space Shuttle Challenger mission found that eggs were formed and fertilized in space. However, many failed to grow past a certain stage of embryo development.

On that same mission, embryos of the stick insect Carausius Morosus suffered significant damage during early development due to weightlessness. The majority didn't hatch. Sadly, the ones that did hatch had short lifespans and showed a large amount of genetic damage such as deformed antennae and abdomens.

The Medaka fish, the first vertebrate to mate in space, had a good success rate with normal development of embryos during space flight. The fact that a vertebrate formed normally may be a good sign. But much more research is needed. Frog eggs for example were found to "not have true radial symmetry" when tilted off axis while maturing, the resulting tadpole had the same "error."

Pregnancy

As I talked about earlier, being pregnant while in space has significant risks to the baby, depending on how long the mother is in space. Since the recommended amount of radiation during pregnancy is 5 rems or less, then a short one or two week trip to low Earth orbit might not be a problem. However, the studies on radiation damage were done primarily here on Earth with x-rays on non-human embryos. Space has different types of radiation exposure and human cells react differently in some cases than other creatures.

Not all radiation effects are the same. "Scientists have been assuming that protons damage cells in a way similar to x-rays, but our results indicate that these assumptions have been wrong," said Lead Researcher Betsy Sutherland with NASA's Space Radiation Lab in a recent Radiation Research article. "The new data show that protons produce more potentially lethal double-strand breaks, a type of severe DNA damage, than other kinds of DNA damages," Sutherland said. "This means that scientists don't really know how human DNA is affected by the most numerous particles in space and, as a result, do not know how to design the proper protection for astronauts."

In the February 2006 edition of the journal Radiation Research, Sutherland and Hada wrote about their research. They found that high-level radiation beams from heavy sources (such as protons) created damaged clusters more than radiation from x-rays (a photon based radiation). Damaged clusters are dangerous because they can cause genetic mutations and cancers, or they can be converted to breaks between the DNA double-strands.

The researchers used beams of high-energy charged particles (protons, as well as iron, carbon, titanium, and silicon ions) and exposed DNA to each type of radiation. They then measured the levels of three kinds of damaged clusters, as well as double-strand breaks produced as a result of the exposure. Because these clusters and double-strand breaks may have different effects on human cells, it is essential to know how many of each kind are produced by the radiation an astronaut or space tourist would encounter.

"Our study shows that we need to re-evaluate the effect of protons on biological systems, even the effects of low-energy protons," Sutherland wrote. "For example, low-energy protons are routinely used in cancer-tumor therapy, but there has been almost no research done on the effect of protons in tumor cells because everyone has assumed that they act similarly to other low LET radiation types, like x-rays. Therefore, this work may help lead to improved cancer therapy."

No astronaut has flown into space while pregnant. None of the space agencies, including NASA, are willing to take that risk. Flight surgeons do a pregnancy test on women astronauts several times leading up to the flight and again just before launch. No woman astronaut is going to miss her flight after years of waiting and training, so I'm sure these women take extra precautions. As Greg Benford wrote in his science fiction book, The Martian Race, "Astronauts do not get pregnant by accident."

There have been no reports of fertility problems with astronauts, but this might be due to medical confidentiality. We know that both men and women who have flown in space have become parents after returning to Earth. Cosmonaut Valentina Tereshkova and Cosmonaut Andrian Nikolayev had both been in space before they had a child together. Astronaut couple Rhea Seddon and Robert "Hoot" Gibson have also both been to space and back several times, and they have three children together. So the flight itself, in these cases at least, doesn't seem to have affected the capability for conception after returning home to Earth. That might not be true for longer journeys outside of low Earth orbit where the total radiation exposure could be significantly higher. But, since no tests have been done, we can only speculate what may happen.

Cosmic radiation, which is always present in space, cannot be fully shielded without a planetary body. Therefore the radiation exposure at a low Earth orbiting space station or hotel during a full term pregnancy would be up to 60 times the recommended maximum allowable dosage! This could possibly lead to miscarriage, birth defects, and retardation.

What about reduced gravity on the Moon or Mars? On these planetary bodies, at least a person could get underground and avoid the radiation exposure issues. But the effects of low gravity on the human body have not been properly studied. Hopefully studies will include the effect of low gravity on pregnancy, muscle tissue loss, and loss of bone mass, since these are huge concerns for long-term exposure to weightlessness. These problems cannot be studied on Earth; they need to be studied in space. But centrifuge testing on the International Space Station, where low gravity animal studies would be done, looks like it will never happen. The next best test facility – on the Moon, is a long way from reality.

In another experiment, for some reason mice embryos in space grew more cells in the part of the brain normally associated with higher thought processes than mice embryos in a similar stage of development on Earth. Before we start to imagine big-brained space people, the fact is that we don't have a clue about what this really means.

Another possible defect in a developing fetus might have to do with the otoliths. Otoliths are in the semicircular canal of the ear and provide necessary sensory input for balance. In weightlessness, the otoliths of a fetus do not seem to develop, leading to balance problems for the offspring when in a gravitational field.

According to a book by Sekulic, Lukac and Naumovic, who base their data on experiments with rats in space, a human fetus should develop normally for the first 21 weeks. Based on these experiments, weightlessness hinders neither conception nor embryogenesis, which is the cellular division that leads up to the embryo. But unlike an astronaut, a human fetus can't exercise, so past the 26th week the lack of gravity would lead to serious birth defects. Without gravity, the muscles in the back and lower extremities of the fetus wouldn't develop, as well as the left ventricle of the heart.

The bottom line is, we don't understand all the risks, but it just isn't worth taking chances with a baby's life. As Dr. Sheryl Bishop told me, "It's very clear that a pregnancy in space without a facility to provide at least, I don't know, X amount of gravity on a regular basis every day for the whole duration of pregnancy, is risky." The research is unclear, but a short duration trip early in the pregnancy *might* be acceptable. What we have no idea of, is how much gravity is needed. Does the Moon or Mars have enough gravity? We just don't know.

Juniper Jairala, Aerospace & Human Factors Engineer discussed human pregnancy in space with me. She said, "At some point we're going to need to find out, instead of being constantly afraid of the consequences. It's going to get harder and harder to screen and restrict these things from happening. Should we take an in-vitro fertilized primate or even human embryo into space?," she asked. Eventually someone, hopefully NASA, will have to investigate human reproduction issues in more detail. Hopefully this will happen in a controlled way, with no suffering.

For long-term settlement in space, we will need some way to protect the mother and fetus from radiation, and provide some gravity for the fetus to develop at least in the second half of the pregnancy. Even ignoring the effects of radiation, which we should be able to shield against with sufficient mass, a baby born without gravity will probably never be able to live in a gravity field. And a baby born in a low gravity field will probably never be able to survive in a higher gravity field, such as on Earth.

All babies are important and deserve the best chance for a happy, healthy life. Everyone born in space should be able to visit or live on Earth if they so choose. We human beings are smart enough and should be able to find a way to make this happen for our children.

Delivery

An experiment conducted onboard the Mir space station in 1998 analyzed two generations of wheat plants. At the time, David Liskowshy, head of NASA's Fundamental Biology program said, "That was the first time we were able to demonstrate that an organism, any organism, was able to reproduce and develop normally through a life cycle."

Shuttle flight STS-70 launched with pregnant rats onboard in July 1995. The rats had uncomplicated, successful deliveries in space. However, the rats required more labor contractions to deliver the babies in weightlessness than did similar rats on Earth where mothers get a gravity assist to deliver the baby.

There were minimal differences in the bone formations of the baby rats, which made no difference after returning to Earth. Baby rats that there born and 'grew up' in space once back on Earth were mostly normal. However, when rolled on their backs while on Earth, they were never able to learn how to roll over and right themselves. Also, many of the rat pups who were born in microgravity had reduced weight and a higher mortality rate when compared to a control group on Earth.

Childbirth is messy, and in weightlessness the mess, particularly liquid is a major issue because it tends to float around. Space Architect John Spencer talked with me about the possibilities, "A birthing chamber, such that a woman is immersed from the torso down could be designed so that there is fresh water circulated through it, so any byproducts of the birthing process is taken out. The water stays clear so the surgeons, midwives, whatever can see the process… these issues can all be worked out," he says.

Caesarean section deliveries are entirely possible and may be necessary in space. Like any surgery in space, blood tends to move away from the body creating great difficulties. The patient needs to be firmly strapped down and the doctor would need to be fixed next to them in order to use both hands. No surgeries have been done in space so far, but astronauts train for the possibility and hope it will never be necessary.

Further Research and the Future

The National Academies, which advises the U.S. federal government on matters of science, set up a committee to review NASA's Bioastrometics Roadmap (BR). In a recent report it said, "The committee notes that the BR contains no references to human sexuality and this oversight should be corrected." Recommendation 2.8 states, "The committee recommends that the issues of human sexuality be addressed in the BR, in relation to long-duration missions such as the proposed Mars design reference mission."

NASA's web site on biomedical research emphasizes, "Ethical, social, and political factors also affect developmental biology, especially in humans. While it might be possible to conceive, give birth to, and rear a human child on a space station, doing so without considerable prior animal experiments would be unacceptable."

So all of the advice that NASA is getting insists that more research be done on reproduction in space, as well as examine the social aspects of sex in space. But NASA seems unwilling to address the matter.

Animal experiments are difficult, because most creatures, mammals especially, are greatly disturbed by weightlessness. Fish, except for the Medaka fish mentioned in Chapter One, swim wildly trying to find which direction is "up." Rats and mice cling to the sides desperately trying to make sense out of their new environment.

The data on humans is even more limited. Most likely people who spend a limited amount of time in low Earth orbit are able to conceive and have normal children after they return home to Earth. We don't know about longer durations, the effects beyond low Earth orbit, or the effects of human pregnancy in space. But all of the research on animals indicates that it would be a very bad idea.

Disturbingly, in a new era of more focused NASA goals to reach the Moon and Mars, the NASA budget has eliminated the majority of biological flight research. Biologists who work on all aspects of health and the human body in space are extremely concerned about the impact of severe cutbacks in this field.

A key question is how much spinning or artificial gravity is needed to keep the human body healthy. Hopefully future long duration flights will use a spinning spacecraft to simulate gravity. This may help prevent some of the long term problems caused by weightlessness. Unfortunately none of the concepts made public so far have shown any spinning spacecraft environment.

Determining how much gravity the human body needs is the key to understanding how the body will change in space and how detrimental effects can be lessened or even reversed.

Currently, it looks like the 2.5 meter Centrifuge Accommodation Module (CAM) that was scheduled to fly to the International Space Station has been canceled due to limited Shuttle flights. This module, or something similar, is desperately needed to ascertain exactly how much downward force (due to the low gravity of the Moon, Mars, or spinning spacecraft) is needed to prevent excessive muscle degeneration and bone loss. As the American Society for Gravitational and Space Biology put it, "The Centrifuge is a unique variable gravity research device: there is simply no way on the ground [on Earth] to determine the long term risks of Lunar and Mars gravity to living systems."

In the event of a space pregnancy, the clear answer is to avoid risks to the mother and particularly the child, and bring them back to Earth immediately. However, on some missions, particularly long duration ones, that might not be possible. And on any mission, given the radiation dosage, by the time you can test for pregnancy there is already a significant chance of birth defects. So might an abortion be prudent? Though the majority of people in the United States generally support abortion, particularly if there are severe health risks, there is a large vocal group actively against abortion and they have a large political presence. Imagine the public outrage against NASA if it called for an abortion in space on a long duration voyage. No one at NASA wanted to comment on this subject for *Sex in Space*.

Another way that reproduction might not be an issue is sterilization, As Dr. Sheryl Bishop told me, "It looks like a pregnancy would be disastrous, for the mother, for the baby, across the board. Since it's only women who can conceive, they may require women to be sterilized before they send them on a long duration mission." But then she added, "Should men also undergo sterilization? A mission to Mars is such a high radiation environment, that the impact on your genetic material and likelihood to pass on birth defects has a great degree of risk." As harsh as it seems, the implication is that we may have to sterilize both men and women who choose to be part of long-duration missions outside of low Earth orbit. Or, perhaps astronauts and space travelers could bank sperm and eggs for use later in life.

Are people willing to be sterilized to potentially be the first people to go to Mars? Clearly there are many who would, especially if they are finished having children, or able to store their sperm and eggs for later. But for those space travelers who do not intend to be the first deep space explorers, sterilization is almost certainly the wrong choice. What about later on when developers and colonizers head into space? It will be a more difficult problem for them, especially if they intend on having children on the Moon or Mars. Hopefully, more research will be done before this happens. With any luck, the results will lead us to a happy solution.

Next, we will consider Sex on the brain and lust in space.

———————————

Chapter 4: Sex on the Brain & Lust in Space

The Sex Drive in Space

Does being in space have an effect on the male or female libido? If so, how might this affect a romantic encounter in zero-g? Let's find out.

The stress of launch, Space Adaptation Syndrome (SAS), and a general flushing of excess bodily fluids can cause a decrease in the sex drive for the first few days of spaceflight. Don't worry, however, the effect is only temporary. Honeymooners and other lovers will still be able to have fabulous sexual encounters in zero-g.

About half of all space travelers experience SAS, so you, your partner, or both, may fall ill for the first few days in space. Medication does help, but the side effect may be a lack of interest in sex. Even if you are not suffering from SAS, or taking medication for it, your body will still need time to adapt to weightlessness. Congested sinuses may lead to headache, and a flushing of excess body fluids will lead you to the zero-g toilet several times. All of this adaptation stuff seems quite *un-sexy*. What are romantic vacationers to do?

After a few days of adaptation to space, the romance returns. The uniqueness of being in zero-g, and seeing the Earth, Moon, and stars from a totally new vantage point, will awaken the libido. The positive physical effects that zero-g has on the way a person looks will make the anticipation of sex overwhelming to vacationing lovers.

Space seems like the fountain of youth. After adaptation, the body is longer and thinner, the wrinkles go away, and breasts perk up. Apparently men have an

added effect. Dennis Tito, the first space tourist to pay his own way into space, reportedly told some friends about this effect, saying that in space there is no need for Viagra. The erection stimulation, or "Space Viagra" effect in men, is purely physical, however, there is no feeling of sexual arousal. But as any man can tell you, a morning erection can easily turn into sexual activity.

Dennis Tito, the world's first space tourist, boards the International Space Station

The next obvious question is; do women experience a similar effect? None of the female astronauts were willing to comment on it for my book. However, Sexologist Ray Noonan, who studied the possibilities of sex in space for his doctoral thesis, gave us a good idea of the answer. He wrote, "Information about whether women have experienced clitoral erections and vaginal lubrication has not been forthcoming, although the homologous nature of this aspect of the sexual response cycle would support the conclusion that they have," he explained.

In other words, since the initial stimulation in men is caused by a higher than normal volume of blood in the penis, and the tissues that make up male and female sexual organs are similar, a lot of blood flowing to the clitoris and surroundings will produce a similar effect in women. Even though women astronauts haven't reported the experience, it has almost certainly happened.

For actual psychological arousal of the libido in space, there is no hard data, but there are some anecdotal stories that suggest a *decrease* in sexual desire. This may be due to the business of the short-term mission, or the isolation experienced on a long-term mission. As discussed in chapter 3, one study suggests that stress-induced hormone changes, specifically a decrease in testosterone production, reduced the sexual drive in men both during, and just after, spaceflight. But testosterone levels returned to normal just one day after returning to Earth.

Testosterone is apparently largely responsible for libido in both males and females. The male hormone testosterone is produced in the testes in men, and is also produced in small amounts by the ovaries in females. It would be dreadful if a honeymooning couple went into space with the intention of experiencing fantastic space sex, but didn't because they just weren't in the mood. The key here may be to book a one-week, or even a ten-day, vacation to space and relax for the first few days while adjusting.

Is the decrease in testosterone purely stress induced? If so, will that level of stress decrease as space travel becomes safer and more routine? We just don't know. If we as a species are to play, live, and work in zero-g, then we must find out exactly what libido changes occur in short and long-term spaceflight. We need to know why this happens and how to fix it. As Ray Noonan wrote, "Data showing decreased sexual drive in men in space underscores the need for such a study." Any volunteers?

Attitudes About Women in Space

The majority of space workers are men, and it has always been this way. This was especially true at the outset of human spaceflight in the 1950s and 1960s when sexist attitudes towards women were prevalent.

Shortly after his historic 1962 orbital flight, NASA Astronaut John Glenn said,

"I think this gets back to the way our social order is organized really. It is just a fact. The men go off and fight the wars, and fly the airplanes, and come back and help design, and build and test them," he said. "The fact that women are not in this field is a fact of our social order."

"I think we all look forward to the time when women will be a part of our space flight team," said NASA Astronaut Training Officer Robert Voss in 1963. "For when this time arrives, it will mean that man will really have found a home in space; woman is the personification of the home." A nice thought, but it still seems a bit sexist to me.

In 1962, Wernher von Braun, chief designer of the U.S. rocket fleet, demeaned women by characterizing them as nothing more than playthings for men. In a reply to a question about female astronauts he said he could, "Reserve 110 pounds of payload for recreational equipment." The remark came off as a joke, but it was really a way for him to brush off the seemingly ridiculous notion, at least to von Braun, that women could be capable astronauts.

Von Braun with President Kennedy

These statements are perfect examples of what most Americans believed at the time; a woman's place was in the home. Apparently, the only way for a woman to get into space was to be the cook and maid in some man's space house. What nonsense!

Astronaut and Biochemist Shannon Lucid remembers these attitudes from when she was growing up. "When I was in school and I wanted to go to college," she told me, "the teacher just laughed and asked 'Why do you want to waste your parent's money, because you're a girl?' When I was studying chemistry, they said, 'there's no point in doing that, go learn how to type so you can get a job.' There was no encouragement for females back in those days." Lucid, however, went on to become a biochemist, one of NASA's first women astronauts, NASA Chief Scientist, and motivator-in-chief of sorts for NASA's Return to Flight program after the 2003 Space Shuttle *Columbia* tragedy.

With the sexual revolution of the late 1960s and the U.S. Civil Rights Act of 1972, attitudes quickly began to change. In the mid 1970s, it was time for NASA to hire astronauts for its new Space Shuttle program. There were seven women within the thirty-five-member astronaut class of 1977, including; Anna Fisher, Shannon Lucid, Judy Resnik, Sally Ride, Rhea Seddon, and Kathy Sullivan. All flew into space on various missions. (For the scoop, read my book *Women Astronauts*, Apogee Books, 2002) Even then, there was an attitude of sexism by many of the male astronauts who were former military. It wasn't until the 1991 Navy Tailhook sexual harassment scandal, however, that the military, especially the Navy, changed its attitude towards women. Since then, military members have been educated about sexual harassment, and at the very least, have been forced to suppress sexist thoughts and actions. Over time, this has probably led to genuine changes in attitude.

Today there are more women who work in the technical fields of space exploration than ever before. But it is still men who hold the vast majority of powerful positions. Somehow, however, either because of society, peer pressure, or just plain sexist attitudes of a few college professors, women are subtly, and sometimes overtly being discouraged from pursuing technical careers. For example, I know of several women astronomers who, after earning their doctoral degrees, have decided to pursue communication careers because the field is much more welcoming to women. Fortunately, not all are discouraged.

The increasing number of women in space career fields continues to change things for the better. My book, *Women of Space: Cool Careers on the Final Frontier* (Apogee Books, 2003) shows the variety of space-related careers that women are pursuing in science, engineering, communications, medicine, and the arts. As in all other human endeavors, diversity leads to positive change. Differing points of view lead to unique solutions, and eventually to the best solution.

Book cover for "*Women of Space: Cool Careers on the Final Frontier*"

"Glass ceilings are not usually obvious anymore," former Jet Propulsion Laboratory Mars Engineer Donna Shirley told me. "How it works is, and I went through this, as you get into management, you're just a lot more comfortable with people who are like you than with people who are not like you," she explained. "When I was the manager on the Mars Exploration Program, I went out, and I think, I was pretty balanced about it, but of course, I was probably just as prejudiced, and I brought in the best people and a lot of them turned out to be women. I had a far higher percentage of women working for me than the men did working for them," she said.

Donna Shirley believes that this isn't really overt discrimination; it's just the way people are. "It goes on not just in engineering, but business, finance, you name it," she explained. "It's just that people are more comfortable with people like themselves. To reach out and really embrace diversity is uncomfortable."

Cultural differences cause *sexual* problems too. Russia appears to be less open to sending women to space. In 2005, Anatoly Grigoryev, the Russian director of the Institute of Medical and Biological Problems (IMBP), declared, "After all, women are fragile and delicate creatures; that is why men should lead the way to distant planets and carry women there in their strong hands." In his opinion, an all male crew should make the first human venture to Mars. His opinion would not be quite so bothersome if he wasn't in the very powerful position of director of the institute that manages cosmonaut crew selection.

Technically, NASA, as an agency of the U.S. government, is not allowed to contract with entities that practice any sort of discrimination. If someone wanted to press the issue, there might be a case for NASA not being able to buy a seat for a U.S. astronaut any longer on *Soyuz* flights, or anything else from the Russian Space Agency, until their discriminatory policy is changed. But realistically, I don't see that happening.

Fortunately, Grigoryev's views are not shared by much of the professional community in Russia. Space Psychologist Sheryl Bishop told me that her Russian science and medical colleagues at the International Space University disagreed with Mr. Grigoryev's statement.

Hopefully, the 'women are weaker than men' attitude does not prevail in the U.S. anymore. Back in 1988, however, Dr. Patricia Santy, a NASA Psychiatrist and Flight Surgeon working at Johnson Space Flight Center said she feared, "The agency may try to evade publicity about space sex by excluding female astronauts from long-term flights." This has not been the case, at least when there have been multiple crewmembers.

So far, no crew has consisted only of one male, and one female crewmember. All two-crewmember missions have been made up of men. There have been three-member crews with one woman and two men, however, on *International Space Station* (ISS) missions. In fact, the three-person *ISS Expedition 14* crew, will include two men, Cosmonaut Mikhail Tyurin, and NASA Astronaut Michael Lopez-Alegria, and one woman, U.S. Navy Commander & NASA Astronaut Sunita Williams.

Lonely in Space: Isolation

Astronauts on long-term missions in Earth orbit must go through several stages of adjustment before truly settling in to life in space.

The first phase of adaptation to life in space is called the "Acute" phase. During this 60-day period, an astronaut adjusts to space both physically and emotionally. The voyage that the astronaut has spent so long training for is new and exciting. At this point, there is a good level of activity.

The "Intermediate" phase is characterized by increasing fatigue and decreased motivation. The monotony of the environment is beginning to take its toll.

Sixty to 90 days into the mission, the "Main" phase begins. Characterized by a lack of productivity, astronauts become irritable and are prone to emotional outbreaks that are out of proportion to the stimulus. Sleeplessness is typical in this phase. Small irritants become hugely upsetting, while good news leads to weeping from happiness. Almost all crew problems that come up happen during this phase. At this point, the smallest disagreement can bring on a big fight with another crewmember. Crewmembers either become very close at this time, or become distant from each other. In this dangerous phase, marriages might be tested, and adultery could be committed.

The "Final" phase happens during the last two to four weeks of the mission. The crew becomes hyperactive as they realize that they will soon be heading home. They try to complete all of their work, especially if they have gotten behind. A delay in the end of the flight can be a significant emotional let down to astronauts anticipating a return to Earth and a reunion with loved ones.

If space travel were less expensive, it would be possible to increase productivity by replacing crews more often, ideally every 90 days for psychological reasons. Without a breakthrough in propulsion technology, however, a 90-day crew rotation for future Mars missions will be impossible.

Men vs. Women: Gender and Isolation

Do men and women perform equally under these circumstances? A 1985 NASA study titled, "Living Aloft: Human Spaceflight Requirements for Extended Spaceflight" by Conners, Harrison, and Akins states, "Women have performed equal or superior to male counterparts in most isolation studies done in Polar situations, underwater habitats, and fallout-shelter studies."

Psychologist Sheryl Bishop warned me about the results of these studies, "You have to be careful about making assertions when you are talking about a lot of small groups," she says, "because no single small group is really representative of any other small group out there." She went on to explain, "We keep finding that mixed gender crews tend to do better in long duration situations."

Why is it that problems occur in the "Main" phase of the mission? In his 1949 book, The Organization of Human Behavior, Donald Hebb hypothesized that the tediousness of the same surroundings and tasks led to deterioration in the ability to think effectively. Not surprisingly, this theory has led to brainwashing techniques. Hebb's later work made a distinction between true sensory deprivation, which is hard to achieve, but can lead to hallucinations, and the effects of monotonous environments such as a space station or jail cell.

Individuals vary greatly in their ability to deal with isolation on long duration missions. But since humans are social animals, groups have a much greater tolerance to isolation than does a single individual. That's why solitary confinement is considered to be such a harsh punishment. A long-term space mission inherently has a strong element of isolation and monotony, which could lead to serious crew problems. It seems that a group of men and women with different personalities would make the happiest crew.

One of the prime sources of stress and burnout in space is an unchanged environment. The monotony of space station and the repetition of daily tasks can easily lead to burnout. Therefore, long-term passengers will need a full and varied agenda, as well as a less utilitarian, more welcoming environment, and the biggest picture window possible, of course. Designers of space hotels will probably take a cue from sci-fi movies and TV shows that portray comforting interior design. Perhaps large video walls with periodically changing images of Earth would help make crews and passengers feel comfortable and less confined.

The Four Horses of Isolation

Isolation and monotony cause four major psychological changes in a long-term crewmember; intellectual impairment, motivational decline, sleeplessness, and severe mood changes.

Both cosmonauts and astronauts have reported intellectual impairment while on long duration missions. Decreased attention and concentration were so severe in some cases, that important changes in the environment, such as a warning light or changes in air quality, went unnoticed.

Motivational decline is the second major change due to isolation. Space missions are risky and when astronauts first come aboard, the rewards seem to be worth the risk. But later in the mission, crewmembers question the rewards of space flight, which seem to have diminished, while the risks remain high. This leads to a 'why bother?' attitude, which can obviously be dangerous.

Sleeplessness is the third difficult change due to isolation, but the reasons this happens are unclear. Lack of sleep can decrease concentration, and the capability to react to emergencies. Sleep impairment seems to be connected to the first stage, intellectual impairment.

The fourth major change due to the isolation of space is change in mood. Over the long months in space, emotions become extreme. Space travelers tend to change quickly from happiness to sadness. Depression is common during this phase.

NASA's first experience with long duration missions began with Astronaut Norman Thagard who joined a Russian crew on the *Mir* space station for several months in 1995. When the Shuttle reached *Mir* to take Thagard home, the event was broadcast live via NASA feed. This proved to be a huge mistake on the part of NASA's public relations staff. As Sheryl Bishop told me, "Norm was gushing away. He was so euphoric. He went on and on talking about how psychologically it had been really tough, and very lonely and isolated. And my mouth was hanging open. 'Oh my, he's being honest. They are going to crucify him.'" She went on to explain, that NASA's public relations staff, "were caught totally off guard, and didn't expect him to let down his guard that much… Sure enough, when they got him back on Earth he went into a black hole. You couldn't see him for weeks. When he finally surfaced, it was all no comment, no discussion on anything he said then. They were trying to do damage control, 'what I really meant was…' And then he very quietly retired from the astronaut corps," she said.

Mixed Gender Crews

So how does sex figure into this environment? Does sex help to ease isolation? Let's take a look at long-duration missions on Earth to get an idea.

Dr. Joanna Wood studies the psychological reactions of long-duration science teams in the Antarctic. She told *Quest Magazine*, "You find yourself down there [in Antarctica] and you have no way out for eight months. So what might have been a passing attraction of mild interest had these two individuals met in the rest of the world, becomes much more intense."

A certain amount of pairing between the men and women takes place. If there are more men than women, as there usually is, then competition takes place among the men who are trying to attract the women. Dr. Wood discussed sexual affairs, "Trying to keep it secret is difficult because it is a small environment," she explained. "When word gets out, there's sometimes some resentment among the other guys who were interested and didn't get a chance. There's strong disapproval if either party is married."

So it appears that an equal number of men and women might be beneficial. As Sheryl Bishop said, "In the end, what we find is that a well balanced group tends to be composed of men and women. And that balance comes from differences in their social capabilities, differences in their personalities, differences in their communication skills. When you have a mix, you are much more likely to have a crew that, in the long run, is better off."

"If there is any changing of partners, it gets even uglier," Dr. Wood explained. This creates even more tension than it might in normal society where people can get away from each other after a falling out.

Astronaut Psychology: What Kind of An Astronaut Are You?

What kind of a person becomes an astronaut? What kind of a person becomes a space tourist? Who are they, what are they like, and what are their attitudes about sex? We'll take a closer look in this section.

Astronauts have been subjected to psychological profiling since the beginning of spaceflight. Mercury Training Manager Dr. Randall Chambers explains how harsh the testing of Mercury astronauts was in his book, *Getting Off The Planet* (Apogee Books 2006). Less extensive tests were conducted on the Gemini, Apollo, and Shuttle era astronauts. Candidates were wary of these profiles and, during interviews, tried to give the answers that he or she imagined the psychologists were looking for, or at least ones that would get them hired. From all evidence, this remains true today.

How important was the psychological profile in selecting astronauts? All indications are that the test results were essentially ignored. According to Dr. Patricia Santy in, *Choosing the Right Stuff, the Psychological selection of Astronauts and Cosmonauts,* other than determining that the astronauts were sane, it doesn't appear that the psychological profile was used at all in the astronaut selection process, or even later in the creation of astronaut crews. Oddly, all of the early data has been destroyed.

Historically, pilots and astronauts have been suspicious of flight surgeons, since these doctors have the power to pull flight ratings and end careers in an instant. In his book *Riding Rockets*, former Astronaut Mike Mullane wrote, "When his stethoscope came to your chest, or that blood pressure needle was bouncing, it was your career on the line." He admits, "I had known pilots who would secretly visit a civilian off-base doctor for some malady rather than bring it to the attention of a flight surgeon."

Astronauts are a strongly competitive bunch. They must be, in order to stand out from the millions of people who would like to go into space. Of that number, tens of thousands of people figure that they have a good shot at becoming an astronaut and fill out dozens of pages of application forms. The selection committee, made up of astronauts and other powerful space agency workers, weed out almost all but a few applicants. In the end, a few hundred people are invited to interview with the committee because they are considered to be the best in their field and in good physical shape. This means that by the time a candidate is interviewed, they have essentially already been selected. So while many can fulfill the basic requirements in education and physical fitness, only those who seem to be the best of the best are chosen. These are the most competitive of the group, and seem, upon interviewing, to stand out from the crowd.

An astronaut, or potential astronaut, must be friendly and personable. As anyone who has been on a job interview knows, you need to impress your interviewer on a personal level to even have a chance at getting the job. Potential astronauts are interviewed by current astronauts who must ask themselves some questions, 'Could I work with this person?' 'Could I rely on them and trust them to keep me alive in a crisis?' If the answer to either of those questions is no, the candidate is sent home. This doesn't seem like an exact science.

The competition doesn't let up even after an applicant has been selected and begins basic ASCAN training. Many of the current astronauts are still awaiting their ride into space, so a new astronaut must prove his or her worth. They must convince other astronauts and key personnel that they can do the job and work as part of a team. As Sheryl Bishop told me, "It's very volatile. It means literally you've got to work to stay on everybody's good side. You get blacklisted by certain people at Johnson [Space Center] and you're never going to fly again," she said.

After selection for a specific mission takes place, the astronaut teams train intensely for at least a year, and usually longer. The training includes long hours and simulations on everything they might encounter during launch, while in orbit, or upon return to Earth. (Everything means everything; there's even a simulated space toilet with lights, camera, and cross hairs so that astronauts can practice positioning themselves right over the target. It's important to minimize the cleanup from a bowel movement in space. Reportedly, hitting the side of the bowl as air sucks waste away leaves a mess that astronauts must wipe up.) Long hours spent away from family causes problems for many astronauts. The divorce rate is quite high in the astronaut corps, even for marriage between astronauts.

Then these socially personable, highly driven, competitive people, who have possibly strained their family lives, have to say goodbye to everyone they know and get on a rocket for a controlled explosion that will send them into space. The chances of death are significant, but according to Sheryl Bishop, "There's not a single person who flew on *Columbia* or *Challenger* that if alive and were offered a chance to fly again, would say the risk was too great."

This same attitude of exploration and a willingness to take risks in order to do something unique may very well translate into astronauts covertly experimenting with sex in space, as discussed in Chapter One, *Has Anyone 'Done It' in Space?* John Spencer, a Space Architect and the President of the Space Tourism Society thinks so, "[Space travel] is a risky business, but that adds some of the mystique to it. When people are involved in risky business they also become more primal too. The more primal you are the more likely you are to want to engage in sex," he says.

The Orgasmic Astronaut

In *The Right Stuff*, Tom Wolfe writes that women threw themselves at NASA's first seven astronauts. So much so, that Astronaut John Glenn is said to have told the others, "keep your peckers stowed."

Nowadays, even though there are many more astronauts, the allure of the spaceman (or spacewoman) is still with us. According to former Astronaut Mike Mullane, "There was an even more powerful pheromone than jet-jockey wings and SEAL insignia; the title *Astronaut*." He wrote in his book, "The fact that none of us had been any closer to space than an airline flight attendant didn't seem to matter. To the space groupies the title was good enough. We males found ourselves surrounded by quivering cupcakes. Some were blatantly on the make, wearing spray-on clothes revealing high-beam nipples, and smiles that screamed, 'Take me!'"

The appeal of being an astronaut wasn't limited to the men. In their astronaut flight suits, "Judy, Rhea, and Anna stole the audience," Mullane wrote about three women in his 1977 astronaut class. "The flight suits seemed to transform them into fantasy creatures like *Barbarella* or *Cat Woman* or *Bat Girl*."

The first American women astronauts, in their cool flight suits. From the left: Shannon Lucid, Rhea Seddon, Kathryn Sullivan, Judy Resnik, Anna Fisher, and Sally Ride

Are things much different today? Probably not; the astronaut is still a rare type of person. He or she was considered to be the best in their field, from their country. They have done something that few other people have done. "Power is the ultimate aphrodisiac," said Henry Kissinger, but the fame that comes from being a superstar, which is what astronauts are, is just as powerful. A friend of mine who once dated an astronaut admitted to me that part of the sexual attraction came from the fact that the object of their affection was an astronaut.

Primal urges drive us to choose the best mate that we can, and astronauts seem to fit the bill. Astronauts are very smart people who are the best in their field. They are physically fit, so much so that a dog breeder would call these types "prime specimens." They are always friendly, charismatic types and were chosen to represent their nation. With all of these things going for them, they have an additional "allure" that comes from doing something that only a few people get to do – fly in space.

I'm not saying that all astronauts are attractive to everyone, because that's certainly not true. But they tend to have a lot going for them and are typically not at a loss for admirers and potential partners; something that I've heard called the "Cult of the Astronaut."

Will this allure last into the era of the space tourist? I'm pretty sure it will, at least for a while. Space tourists are, and will be people who have become successful enough to afford a trip to space. Sometime this century, hundreds of space workers will head to orbital stations, the Moon, and Mars. Even when thousands, or tens of thousands of people go into space, they will still be an elite group, compared to the billions of people on Earth, and the millions who want to go into space. So the appeal of the astronaut, by which I mean anyone who has been into space, may be with us for a long time.

Space tourists will not need to be the best of the best, but they will still be somewhat special, at least for the foreseeable future. These people will continue to be competitive risk takers, because it will be a long time before spaceflight is as safe as travel on airliners. They will not have to be much of a team player, however, because they will pay a team of hoteliers to do the work of getting them to and from their destination. And they will not have to deal with the problems that occur on long-duration space missions, at least not for some time to come.

Sex at NASA

Spaceflight has always been a hazardous business, so the focus has been on the technology and hardware of the missions. Very little time and attention has been given to the psychological aspects of a space mission. That's the correct emphasis, but it doesn't mean that psychological and sexual aspects should be ignored either. NASA's position is that the astronauts are trained professionals, they are up there to do a job and there is no time for coddling people with psychological issues, or for discussing sex. NASA and the other space agencies don't even want to discuss the *possibilities* of sex in space.

"When NASA is pressed for information on how it is addressing the issue of sexuality in a space environment, science suffers," wrote Sexologist Ray Noonan in his 1997 doctoral thesis. "The role of the space scientist becomes mired in a morass of imperatives designed to uphold society's prevalent negative moral attitudes toward sexuality," he wrote.

This is much the same response that medical doctors William Masters and

Virginia Johnson received when they published their Earth shattering studies on human sexuality in 1966. *Human Sexual Response* was harshly criticized by many of the duo's colleagues because they considered sexuality to be an improper field of physiological study. Had the work been on skin diseases or anything else, there would have been no issue. But even today, people are still squeamish about sexuality and are concerned that it is not a "proper" field of research.

I originally tried to work with NASA's public relations staff to get astronauts, flight surgeons, or anyone else who deals with human spaceflight to talk with me on the record for *Sex in Space*. Sheryl Bishop advised me, "If you try to go though the agency, you're never going to get past the front door. Nobody from the official side is going to discuss it," she said. "It's a deep cultural discomfort zone."

That's exactly what happened! When I worked on my previous books, *Women Astronauts* (Apogee Books, 2002), and *Women of Space: Cool Careers on the Final Frontier* (Apogee Books, 2003), I was able to get interviews with astronauts and other people through official channels. I was even able to discuss menstrual cycles and sexism with several women astronauts. But for *Sex in Space*, no one was willing to talk with me. In fact, many NASA workers, who will remain anonymous, played dumb with me, even when I explained that the book would balance the giggle factor with serious scientific questions about reproduction and the future of humans in space. I have come to understand, however that NASA's attitude towards sex reflects the illogical prevailing American attitude about sex.

Sheryl Bishop explained, "I think it's predominantly an American puritanical attitude towards anything that has to do with sexuality. We just don't talk about that. Look at the U.S. military policy 'don't ask don't tell'... It's ludicrous," she told me. But at the same time, our media is filled with sexual content, which is often very overt. Dr. Bishop continued, "It's because we repress it, or it's taboo to talk about it in any kind of profession or recognize it as a normal part of life. Because we are forbidden to talk about it in those venues, we end up putting it into our entertainment to overcompensate for it," she says.

In his thesis, Ray Noonan wrote, "During my research for this study, some of the space scientists with whom I spoke suggested that if I ever hope to get a job in aerospace, then I should abandon sexology as my focus. A few mentioned the fears of colleagues who were afraid of losing their jobs for discussing the topic of sexuality at all, particularly with the press who occasionally would ask about it, or knew of colleagues who were warned not to talk about sexual issues."

Similar problems face many of the scientists at NASA. According to Noonan, "Unconfirmed anecdotal stories, often told by space scientists suggested that they would face real danger of losing their jobs if they discussed the subject [of sex]."

NASA is a publicly funded organization that is, let's face it, optional. Congress can significantly cut, or increase, NASA funding at any time. Since this is the case, then NASA is not going to do anything that would be objectionable to a significant number of taxpayers and voters. And there are a large number of people in the U.S. who do not want government agencies spending money on sex research and reproductive biology. Therefore NASA wants to avoid that controversy and potential loss of public support, so they do not talk about sex in space, at all.

Since John Spencer is a strong advocate of private space tourism, he isn't concerned with prudishness at NASA. "Sex in space is part of exploration. It's an exciting part I think. If NASA doesn't want to deal with it, no big deal we've got other people working on it."

Every time the media raises the topic of sex, NASA gives one of two predictable responses. First, it declares that sex is not an issue at NASA. Second, when pressed that sex will become an issue on long-term space missions, the agency says that sex is not something NASA needs to focus on right now. "It's an illogical position, says Bishop, "but it is the one they are taking."

"I think they're in the 1950s mentality." Rod Rodenberry told me. "My father [*Star Trek* Creator Gene Roddenberry] had to deal with that on TV – no belly buttons. It's alright guys, it's the 21st Century," laughs Rod. "You can have sex with people, even premarital, even fellatio."

In 1985, Dr. Yvonne Clearwater of NASA wrote in a sidebar of a *Psychology Today* article, "It seems obvious… that a group of normal healthy professionals will probably possess normal healthy sexual appetites. In space regardless of crew gender composition, if we lock people up for 90-day periods, we must plan for the possibility of intimate behavior," she wrote. "Our job is not to serve as judges of morality but to support people in living as comfortably and normally as possible while doing extremely important work. This means, as a minimum, ensuring that the environment does not disrupt their well-being, and work performance needs for auditory and visual privacy are met."

Based on the reaction to her sidebar you would have thought she was advocating genocide. It almost killed her career, and she had to stay silent on the matter for a while as NASA public relations dealt with the matter. Reportedly, it took a lot to calm the Congress down after constituents objected to the psychologist's comments. This particular incident has made NASA very skittish at the very mention of sex, and that condition hasn't changed in two decades.

Yet NASA is reasonably wary of the unusually strong attachments and emotional baggage that astronauts who are lovers might carry with them on a space mission. Crewmembers probably wouldn't show favoritism, but might it be *perceived* that way? Or if a relationship goes sour in flight then there is no way to escape. This is why the agency has avoided the issue of couples in space flight, and why it will almost certainly continue to do so.

If partners are separated on long voyages, then might one of them (in space or on Earth) stray and have a romantic and/or sexual affair? These extramarital relationships, if they were discovered, could undermine existing friendships and cause severe problems and distractions for the astronaut on the mission. Imagine the effect of getting a 'Dear John' or 'Dear Jane' letter while on a 22-month Mars mission. The results could be disastrous.

It's hard to know which is better for crew stability; having astronaut couples or not. NASA's approach to congressional appropriations is not to annoy any large segment of the population which might be disturbed by sex. Since a large portion of the American public would be upset if it found out that there was sex in space, then it is best not to mention it. This is true for both scandalous extramarital affairs and marital relations. People would ask, 'Why should the taxpayer foot the bill for one couple's dream honeymoon?'

It seems that we probably won't hear space travelers talk openly about sex in space until we have true space tourism and couples are paying for their own romantic vacations in orbit. Western society in particular has difficulty accepting sexuality as a healthy part of normal life.

"It's an interesting denial that we are in," says Sheryl Bishop, NASA, "will theoretically and intellectually agree that [sex] has the potential for being an issue. However, it's going to be a bridge they cross when they get there. So until they have to select a crew for a long duration mission, they are not going to deal with it."

In "Living aloft: Human Requirements for Extended Spaceflight" a 1985 NASA publication by Conners, Harrison and Akins, the writers contend that psychological and social factors will become, "Increasingly important determinants of the success or failure of future space missions." So, it's something that we will have to address eventually. Hopefully the first time we address it will not be when a space traveler gets pregnant during a long-duration mission!

Samuel Coniglio, Photographer and Vice President of the Space Tourism Society expressed these thoughts on NASA's attitude, "Romance, intimacy, and sex are as natural as breathing. NASA's politically safe approach of ignoring sex is fine for a government organization focused on science. Business though is another story. Sex is the biggest business on the Internet, and has always been a motivator for new technology and new industries. The time has come to get out of the closet and start businesses that promote romance in the final frontier," he says. There will be plenty of customers." I'm betting that Coniglio is right, because you are reading this book.

Gender, Sex and Crew makeup for Long Duration Missions

Currently crews stay on the *International Space Station* for 6 to 9 months. The longest anyone has lived in orbit is about 430 days. Antarctic research camp experiences are always a year or less. But current plans for a Moon base and an eventual Mars mission will take two years or more away from the Earth. This is, of course, unless there is some huge technological breakthrough in propulsion that can get people to Mars much more quickly.

A mission to Mars, with a previous Moon mission to verify the capability of the hardware, will require a new level of social stability and psychological challenges in excess of anything that has been attempted to date. What kind of crew makeup is needed, and will the crew be sexually active?

Historically, in any long-term remote expedition or exploration into the frontier, alcohol has been used as a recreational drug to relieve stress. This has been true on long sailing ship voyages, long overland journeys, and even long dog sledding treks. Would social drinking be allowed as a release on a long space flight to Mars? Maybe. The Russians allow drinking in space in limited quantities, but NASA rules would have to change. If an astronaut were to get really drunk in space, he or she might get everyone killed if not careful; there are a lot more hazards in a spaceship. Hopefully any intoxication would be kept under control. The crew would need to agree to

Mars as photographed by the Mars Exploration Rover Navigation Camera during approach

some self imposed drinking limits, because there would be no way for Earthbound flight controllers to enforce drinking limits.

Boredom on long space flights is a constant enemy. To fight this, the crew would have to be very interested in their activities, and the food would have to change constantly. Food, by the way, is a very difficult problem when you have to pack years ahead of time and only have limited capability to perhaps grow a few plants. Even the interior of the spaceship would need to be carefully tailored to raise visual stimulation among the astronauts. Perhaps high-definition video walls would help to give the astronauts a sense of variety.

Sex is natural. It is as natural as eating, sleeping, and defecating. Not all of those are things that we want to do in public, or watch, but we don't try to hold them back indefinitely either. To propose a sexless voyage that lasts for years is simply unrealistic. Studies of prisons and long ship voyages show that even normally heterosexual men and women will turn to homosexual relations in order to purge sexual urges and relieve boredom and tension.

Sex should not be suppressed especially since medically we can prevent pregnancies and sexually transmitted diseases with proper screening and medical precautions. Asking people to refrain from sex when on long duration missions is unrealistic. And there would be no way to enforce it once the crew was underway. If they want to, members of long duration crews should be able to have sex, without someone else making a big deal about it. And if they choose to refrain, or limit it to self-stimulation, that is also their choice to make.

Dr. J. Annexstead of NASA's Johnson Space Center and eleven time Arctic expedition member and two time expedition leader noted in *Quest Magazine* that in Antarctic crews with women, there seems to be less competition and crews seem to get along better. Women in isolated crews tend to perform a socializing function as well as their mission function.

I've watched my husband interact with his friends when gathered together, and have come to the conclusion that there is more competition when men are alone than when men and women are in a group together. I think a mixed crew is more likely to be calmer, but any group can have personality conflicts. Remember Sheryl Bishop's earlier comment, "You have to be careful about making assertions when you are talking about a lot of small groups, because no single small group is really representative of any other small group out there."

And Dr. Bishop stated the most important aspect of crew harmony, "When you have a small group, if the focus isn't on keeping the entire group together as a whole, as a team, where everybody shares and everybody participates and everybody has a say and can be heard, at any point in time that people start sub-grouping, whatever causes that subgroup to form, then you start having an issue."

Carol Ellison, a psychologist specializing in sexuality, told *New Scientist Magazine* that, "Break ups can lead to violence and all kinds of things. People are very primitive in their emotions around partnering and sex." She went on to say that sex on long missions, "could help or hinder, depending on how many people you've got, their relationships, and what it means to them."

Solar System Ambassador Ginny Mauldin-Kinney feels that relationship problems such as breakups will happen no matter how much we try to prepare crews for any situation. She told me, "The potential for these scenarios would arise over time no matter how hard we try to stifle them; perhaps along with the urge to shove someone out of the airlock due to jealously! Seriously, biological urges and human emotions will need to be taken into consideration. Psychological profiling of long-term space travel will require massive study" she says, "It is important to understand that no matter how hard we try, the study of the human psyche is not an exact science."

So when we select a future long duration crew, I would hope that we follow

Dr. Bishop's advice, "You choose the emotionally mature individuals. You give them lots of training in conflict resolution. You give them lots of training in talking about these issues [like sex] before they go. You put it [sex] on the table and have it discussed. You make them aware that there are likely to be attractions and then as those attractions wane, disinterest. How are they going to handle that? And you let the group deal with it ahead of time. You don't deny that it's going to happen and you certainly don't shove it back out of the way because it's uncomfortable," she says.

Is it really necessary to have couples in long duration missions? Maybe not, but Dr. Bishop pointed out at least one study that showed it was helpful, "Gloria Leon ran a study of three couples that were intentionally icebound. They did rather well, but each one of the couples said, 'If I hadn't had my wife or husband to talk to it would have been unendurable.'" It seems that bringing your soul mate along for the ride may be good for business.

Other studies of Antarctic expeditions show that married couples handle the isolation better than single people. But singles in the group might resent the couples and feel isolated and lonely. And there might be a perception that two people who were in a couple would be more loyal and supportive of each other than other members of the crew. Even if this wasn't the reality, even the *perception* of unfair treatment would be enough to disrupt crew teamwork.

Sometimes personalities clash, and sometimes they mesh. For long term missions it is extremely important that the crew get along very well as their friendship will be tested during the mission and any disagreements will compromise mission effectiveness. If possible, it would be a good idea for everyone to spend several months in simulated mission conditions in a remote location. However, this isn't normally done, and might be regarded by the crew as a waste of time.

Exploring Mars.
Artwork by Pat Rawlings, Courtesy of NASA

Dr. Bishop's suggestion for crew selection testing was, "I think a winter over in Antarctica, nine months, would be an absolute minimum. You send down a larger pool of people and out of that larger pool you select the people who will go, and the backup crew. That's a realistic way to maximize your chances to get a group that's well integrated and likely to be able to do a three-year mission."

Author and former NASA Flight Controller Marianne Dyson explained her fears about sexism and crew selection, "My primary concern about Mars missions is that women will be barred from going because there is not enough medical data to show they will not get breast cancer, or become infertile, or suffer other serious health problems as a result," she told me. "The *ISS* could be used to gather this data, but the Russians are only sending men [to the *ISS*]. And the American astronaut corps is fixed at 20 percent female and minorities. I've also been told that data is not collected by gender to protect privacy!"

The obvious solution is to send more women, stop being squeamish, and start collecting real science data on radiation exposure, breast cancer, menstrual cycles, and female fertility. This is a case where NASA attitudes on sex might lead to discrimination, and even more importantly potentially causing (or not preventing) real potential health problems.

The Love Boat

My first thought was that since couples help each other in isolation situations, then we should send married couples. But Dr. Bishop pointed out some possible flaws with that situation. We discussed a hypothetical case with three couples on a crew of six to Mars where the only people they see are other married crew. "Of the other five people you have to interact with," she explained, "one [person] you have been with for a long time, so he or she gets to be boring really fast because you know everything there is to know about that person. So now the only other four people that you have to interact with, only two are of the opposite gender. I would submit to you that it would be a very rare group that you could send six people out, three couples, on a three-year trip and have them stay totally monogamous."

If true, then an adulterous affair among the crewmembers will become a difficult situation very quickly with resentment and anger, leading to a possible collapse of the mission. Talk about dangerous liaisons. A situation like this could endanger the lives of the crew.

So what is the ideal crew? Dr. Bishop says, "My belief is that whatever the number is, choose all single individuals that are emotionally and psychologically mature, and obviously have the skill sets. We have time to put them together as a group and train them as a group so that they form a group identity, coherence and loyalty to the group early on. They have to leave their family stuff behind even before they leave Earth. That takes some very dedicated people. It's going to be a challenge." She then goes on to joke, "I say jokingly, that an ideal group should have equal number of men and women and they should all be bisexual so that they have double the partners."

The Biosphere 2 project was supposed to be a self-sustaining isolation environment in the desert near Tucson, Arizona lasting over two years. It was a mission to simulate a future space habitat. It's interesting to note that four unmarried men and four unmarried women were chosen for the study – the same balance as Sheryl Bishop suggested.

The sexual activities of the Biosphere 2 crew were not reported because the managers of the project wanted the private lives of the crew to stay private. However, it is reasonable to assume that couples did have sexual relations. In his thesis, Ray Noonan reports that a study on space analogues stated, "In fact, sexual activities did occur in Biosphere 2, although no pregnancies occurred. They said the female crewmembers were advised that pregnancy during their stay would be cause for removal from the project. As a result, the women took it upon themselves to prevent pregnancy."

Future Space Culture

What physical and cultural changes will future space habitation bring to humans? It is difficult to predict, and will depend on the makeup, culture, and habits of the first inhabitants. But if the *ISS* is any indication, there may be less nationalism and a recognition that we are all from Earth. What will the sexual culture of a future space habitat be like? Will it be the same as Earth or will the realities of the situation bring major changes? Science Fiction Writer Robert Heinlein thought that monogamous relationships would give way to group marriages to increase genetic diversity and enable a close knit care giving group to take care of the family's children. Just wild speculation? It's impossible to say, but interesting to consider.

Here are some other speculations, starting with my own. I think that the people in a space habitat of the future will be highly educated professionals who tend to be more open to diversity. They will also each be busy because labor will be at a premium, so having one of the parents stay at home to raise children will be an impossible luxury. Childcare 24 hours a day will have to be available to allow adults to work shifts all day and night. Childcare will probably be a necessity that starts just a few months after birth. Everyone will help to raise the children, and school will probably start very early in life, with lots of time for each child to learn by playing and exploring what they are interested in.

I think that people will still be sexually monogamous for the most part. That may just be my own bias – but I think most people are monogamous and thousands of years of primitive instinct will not be overridden by 'logical' arguments to the contrary. Jealousy is just too strong an emotion to suppress with logic. But there will have to be a greater openness to discuss sexual relations than exists today. No longer will sexual behavior be seen as inappropriate except for procreation. But sexual exploration before marriage, homosexuality, bisexuality, and masturbation will not be the taboo subjects they are today.

A less savory side issue is prostitution. In the past, wherever people have set up such colonies, prostitution has followed. When people go into space to build an orbital habitat, a Moon colony, or something similar, it is probable that people will pay for sex. It may sound sexist, but this is particularly likely if there are a lot of men on the job site and very few women, and if workers are away from Earth for more than a few weeks.

Other people that I talked with for *Sex in Space* gave unique perspectives on future culture. "I keep saying that humanity should unite," Rod Roddenberry told me. "The idea of *Star Trek* is the united federation of planets. But then Arthur C. Clarke pointed out to me that once we get out into space, people are actually going to separate and go their own ways [based on religious beliefs]. And I think to some degree we will populate the solar system, galaxy, and Universe with different species almost. We will evolve differently, culturally, and physically. And then even far in the future we will re-discover ancestors of ourselves who look completely different. We definitely will separate, but it would be nice if we can still work together," he says.

Samuel Coniglio told me, "Given time and with the support of technology, humans will adapt to a place like Mars. Terraforming, the transformation of a planet to an Earth-like environment, begins with the planting of the first high altitude lichen or moss. It may take many generations, but Mars can become more hospitable to humans, and the human body will adapt as much as possible to the harsh Martian climate," he said. "We will become the first Martians."

Author and Space Advocate Vanna Bonta gave me a wonderful description of a future space culture where form follows function, and that changes *who* we are. "Human forms are perpetuated through sex, and sex also perpetuates human consciousness," she explains. "It's interesting to speculate consciousness existing in any number of organisms that may, in the future, be different from what we now recognize as human; subtle or drastic, variations would develop in response to surviving various environments."

"Space settlements would also contain biospheres replicating Earth conditions and atmosphere," says Bonta, "and changes might be less physical but more along the lines of human consciousness."

She continues, "If form follows function, as we know it does in this Universe, then consciousness will adapt to whatever form it requires in order to function. Hopefully, it will also develop its fundamental function; what that is may be debatable within many schools of thought, but it is indisputable that evolved thinking recognizes the universality of Life."

Without a doubt, any future space culture will be something new. But we will take what we know with us wherever we go. Our sexuality will come with us and it will help us to settle the galaxy. Human beings will go to infinity, and beyond.

Chapter 5: To Infinity and Beyond! The Future of Sex in Space

Sex will help us to settle our galaxy, and explore our Universe. Throughout human history, our sexuality has always been a part of exploring and opening new frontiers. We are sexual creatures, and we're lucky that we evolved this way.

Without our sexuality, we would not have been able to settle, grow, and develop in nearly every place on Earth. Instead, our species would be confined to some small part of Africa where humanity first started, or more likely, humanity would have died out there. Space is the same concept, but on a much grander scale. It is our nature to go out into the Universe, explore, settle, and have children if we choose. To be confined to this planet would be tremendously limiting, and would undoubtedly eventually lead to the death of our species. Fortunately, the future looks bright for our expansion into the Universe. We live in an exciting time! We are privileged to be present at the beginning of the age of space travel. We are about to return to the Universe from which we came.

Space Tourism: The Ultimate Thrill Ride

People have always dreamed of flying to the stars, but the true beginnings of space tourism began with the technical fantasies of visionaries like Jules Verne. Technology has fueled our dreams of space flight. Over time, books, radio, movies, television, and the Internet, have created a demand of sorts, for the real thing. Now, with the success of the X-prize in October 2004, there has been an explosion of interest among the general public, and a surge of hope among space buffs, that there may soon be true space tourism for the masses.

"I realized that space tourism is a little bit like sex; it's exciting, it's a little bit dangerous, or risky, perhaps, and when it's well done it's tremendously satisfying," explained former Astronaut Rusty Schweickart at a recent luncheon at the International Space Development Conference. He continued, "But there's more to it than that because, at a deeper level, it's also the basis for reproduction, which is the foundation of survival." He talked about the importance of defending our civilization and planet from asteroid disasters and getting people into space. "It depends on the evolution of space tourism, which reflects the desire of everybody in this room, in one way or another, to get out into space," he said.

Space tourism experiences fall into three categories, orbital, sub-orbital, and Earthside. "Tourism" implies traveling to an experience, so "space tourism" means going on vacation to experience space. The purveyors of space tourism are selling the *space experience*. As John Spencer, Founder and President of the Space Tourism Society and author of *Space Tourism: Do You Want to Go?* told me, "None of us are in the space business. We are in the *experience* business."

While climbing onboard a spaceship and blasting off into orbit is the ultimate way to experience space, most of us cannot afford that option. Not yet, anyhow. Luckily, there are many ways to simulate the space experience without leaving the Earth. Let's explore our options, starting with the ultimate.

Orbital space tourism is the most adventurous space experience you can purchase today. There are very few flights available and only the wealthiest Earthlings need apply. If you can get the approval of NASA and the Russian Space Agency, have about $20 million to spend, and are willing to spend six months in flight preparation in Star City, Russia, then you can spend several days on the *International Space Station* (*ISS*).

The International Space Station

Three space tourists, Dennis Tito, Mark Shuttleworth, and Greg Olsen, have funded their own adventures to the *ISS* via *Soyuz* launch from Star City, Russia. Each hired the company Space Adventures to help arrange their space flight. More space tourists are on the way, including the first official female space tourist, Ansari X-Prize Philanthropist Anousheh Ansari, who plans to blast off to the *ISS* sometime in the near future. She is the backup for Japanese entrepreneur Daisuke "Dice-K" Enomoto, who is scheduled to fly to the *ISS* in late 2006. Businessman Charles Simonyi is scheduled to make his trek to the *ISS* in early 2007. We are still awaiting the first "space tourist couple" to buy seats for their own, very personal spaceflight.

When will orbital tourism become romance and sex in space? "If you had enough money, you could pay the Russians to send two people to the International

Space Station right now," says John Spencer. "You could make a deal with the Russians to have privacy in the Russian module. And they could, next week, have a honeymoon in space. There is virtually nothing stopping people from doing it right now, other than scheduling issues," he explains.

The Space Tourism Society was created to help pioneer the frontier of space tourism. "Our goal is to try to get as many people as possible, as soon as possible, to have the space tourism experience," says Spencer. "But since we can't send a lot of people up to space, at least not yet, how do you experience space? Even for the next 50 or 60 years, there are very few people going off world for real, but millions and millions having Earth based space tourism experiences," he explains. "As our immersive multiple day simulations evolve, more and more people will have that kind of experience."

One way to briefly experience space is a sub-orbital flight, which takes you into space for several minutes on a ballistic trajectory before dropping you back to Earth. In the near future, if you happen to have an extra $200,000 floating around, a sub-orbital flight to the edge of space may just light your fire. Sub-orbital flight involves an extremely expensive, extremely short trip to the edge of space where one can see the blue glow of the Earth's atmosphere, see the stars above during the daylight, see the Earth far below, and experience weightlessness.

Building on their X-Prize success with *SpaceShipOne*, Scaled Composites has agreed to build a commercial vehicle called *Enterprise* for Virgin Galactic. The company, created by the Virgin Group Limited's Richard Branson, will offer sub-orbital flights into space starting in 2008 or so, and plans to someday ferry tourists to and from orbiting hotels. Hundreds of people have reportedly contacted the company proclaiming their interest in putting down a $20,000 deposit on the $200,000 price of a sub-orbital seat. And nearly one hundred people, from at least eighteen countries, have plunked down the entire $200,000 ticket price to become a founding member and passenger.

The first honeymoon in space has already been booked! Space Advocate and Yuri's Night Co-Creator Loretta Hidalgo and her fiancé, National Space Society Executive Director George Whitesides, have booked a sub-orbital honeymoon flight with Virgin Galactic. When the flight takes off in 2008, the duo will become the first married couple to venture into space on a private flight.

Another way to experience what space might be like is to feel weightlessness for short periods of time on a ballistic aircraft ride. Space Adventures arranges flights on Russian aircraft out of Star City, Russia, while the Zero-g Corporation provides this service on the *G-Force One* airplane out of Florida.

For $3700 per person, per flight, Zero-g customers can experience zero gravity as well as the gravity of the Moon and Mars. "Float, flip, and fly the same way orbiting astronauts do," says the web site. The package also includes instruction from real astronauts as to what it's like to float in zero gravity. After the flight, clients are treated to a party, award ceremony, dinner, and Zero-g promotional goodies.

"Experiencing weightlessness is something so unique and exciting that there is a large market for it amongst the general public," Zero-g Corporation Chief Marketing Officer Noah McMahon told me. "We feel that there is a very large market for weddings and honeymoons in weightlessness." My research says he's right on target.

As part of her *Out-of-This-World* contest, Martha Stewart recently showed her studio audience a short video clip of her Zero-g experience. Stewart and friends, including Businessman Charles Simonyi, who has bought a ticket to the International Space Station, flip, float, and dance on this ballistic airplane ride. As the video ran, Stewart narrated, "Oh, there we are dancing. Charles is a great dancer, he's never danced quite like that before." And then the audience laughed hysterically as she fell on top of him.

With advances in simulators today, space *can* be experienced without a rocket or airplane, and for much less money. The Earthside market includes several space-themed immersive experiences such as Disneyworld's Mission:SPACE and Japan's Space World theme park.

Disney advertises its Mission:SPACE ride as, "As close as you can get to blasting off into space without leaving Earth." Tragically, several passengers have died in separate instances, mostly due to preexisting medical conditions. Disney says that Mission:SPACE has been made safer and less vigorous.

John Spencer, who helped design Japan's Space World theme park, has formed Red Planet Ventures, to help people *feel* the future. "We're going to create immersive experiences that you physically go to and spend two or three days, or a week, immersed in that environment that is all space, Mars, future oriented. We're going to take you into the future, he explains."

Tourism is one of the biggest businesses in the world, adding up to more than $400 billion per year in the United States alone. It is not unreasonable to assume that if more people could afford to fly into space, they would, and market studies verify this. A 1993 survey in Japan found that 70% of those polled are interested in traveling into space, and half of people surveyed were willing to pay three months salary to do it! A similar United States survey taken in 1997 by Yesiawich/Pepperdine/Brown of Florida, and Yankelovich Partners of Connecticut, polled 1500 families about their interest in taking a trip to space. The poll concluded that 42% of Americans were interested in going into space, if they could afford it. So, lots of couples are ready to head to space, if the price is right, and possibly experience lovemaking in orbit.

Drawing showing the plans for Mars Base One and immersive
simulation of a second generation base on Mars

"Zero-g is providing the public with its first taste of space through weight-lessness available to everyone," says Noah McMahon, "but the space frontier is being opened wider and someday weddings and honeymoons in space will be commonplace."

Imagine how romantic a space ceremony would be. You would face your favorite person in the Universe with the stars and Earth sparkling behind him or her. This sounds just heavenly to me! Afterwards, you would float to your honeymoon suite and have your own private celebration.

Space Performance Artist Lorelei Lisowsky told me about a 'Moonkiss' performance she did as part of a student project onboard a NASA KC-135 airplane. "I positioned myself next to him, high on the drugs NASA had administered me," she explained, "high from the floating experience, and our lips met. We held there, released and tried again, by now we were in Moon gravity parabolas and felt like we were experiencing the first kiss on the Moon. I felt I had achieved what I had come to NASA to do, feminized the environment even for a moment, with my action, put an imprint within the militarized space," Lisowsky says.

Lovemaking in the weightlessness of space is a unique experience that is not available anywhere else. It's *the* unique honeymoon experience, whether you're just married, celebrating a special anniversary, or just traveling with a good friend who wants to experience it all with you. People are inspired by new sensations, and the space tourism business is all about giving people unique experiences. Sex in space opens up a whole new world for space tourists. Sex in space is the "killer app" of space tourism.

Recently I spoke with Eugene "Rod" Roddenberry, who is working on a documentary about his father and *Star Trek*. He told me, "Art is what inspires people. *Star Trek* was art that inspired people to go to space... Sex in space is going to inspire people, so it's art... It's a part of humanity, it's important." Envisioning ourselves in a new, unique place is what motivates us to want to go into space.

One person's pornography is another person's inspiring erotica. But whatever you call it, it's about a $10 billion a year industry. Whether you like it or not, someday someone in the adult entertainment industry with the money and the inspiration to do something unique will make a porno movie in space. My bet is that the DVD will sell very well because it will be unique. This is also a form of space tourism, because it will bring part of the space experience down to people on Earth. Maybe one day there will even be an XXX-Prize for unique sexual positions and situations in space.

The space tourism community welcomes everyone no matter what gender, ethnicity, sexual persuasion, or age. What matters is that you have an interest in unique experiences that will become more accessible to everyone in the not-so-distant future. We are different, but share the same dream of a brighter tomorrow, and space can make it happen.

Space tourism is, in some ways, a sneaky way to make the future happen the way we want. As John Spencer explained, "The whole thing about space tourism is it's a means to an end. By having a vibrant, healthy, profitable space tourism industry, we're developing a very robust technical and operational capability for off world stuff that will significantly stimulate exploration, utilization, and colonization," he says.

What's happening now?

It's clear that up to now most people have been observers in the space program, because only governments have been able to pay for space exploration. Sure, more people are involved in commercial satellites, but that's not breaking any new ground in terms of getting people into space. Therefore, even people interested in space exploration are unlikely to get a job that directly supports doing something to help the long-term settlement of space.

This lack of direct involvement has hurt the development of space by making the general public apathetic about space. Hopefully that will change as the cost of getting people and things into space decreases. If we had cheap commercial access to space, then we would not need to rely on government programs and could develop the space frontier much faster. Any commercial venture would be able to work towards developing its own space program.

Imagine how slow settlement and development of the Western United States would have been if every trip west of the Mississippi river had to be funded by the government. Two centuries after the Lewis and Clark expedition, we might only have one city and a few small towns out West. The U.S. government would probably still be arguing about whether to start funding the settlement of Denver or Dallas.

The key to opening up space settlement is to make space exploration affordable to many more people and groups. This will allow commercial enterprises to flourish, and people to take chances and spend money on projects that may flop, or may pay off well in the long run.

But is cheap commercial access to space possible? It will be someday, but it's a question of when. Technology is a major factor in the high cost. It's difficult to get a vehicle above the atmosphere fast enough to achieve orbit. With chemical engines and staging, propellant still makes up about 85% to 90% of the total mass at launch. That's simply based on the energy capability of the chemical reaction, and the physics of achieving orbit.

Even if the technology becomes cheaper, it turns out that the launch costs for an orbital vehicle are probably going to remain high for some time to come. Even the most optimistic estimate, shared by Dr. David Livingston and colleagues in a recent seminar at the University of North Dakota titled, "The Challenge of Cheap Orbital Access," estimated that a near future Expendable Launch Vehicle (ELV) would cost about $2000 per pound to operate. That estimate assumes that companies can significantly reduce the development cost and amortize it over many launches. Currently, launch range costs and insurance costs are a significant factor. But this may change in the future. If launches failed much less frequently, and there were a lot more launches, then the insurance and range costs could also go down.

Since the 1990s there have been several ambitious companies determined to make cheap launch vehicles for many purposes, including space tourism. Many of these companies rose up and disappeared, like Beale Aerospace. Others, like Kistler Aerospace, had to declare bankruptcy but have emerged and are trying again. In Kistler's case the company chose to partner with Rocketplane Incorporated. Some, like SpaceX with its *Falcon* launch vehicle, have kept going even after an initial launch failure, but may very well achieve their goals and significantly reduce launch costs.

The X-Prize competition has stimulated a number of new companies to develop cheaper launch access. Burt Rutan, founder of Scaled Composites, and designer of *SpaceShipOne*, has his eye on building a low cost orbital vehicle based on what his team learned while building *SpaceShipOne*. They plan to apply this technology to create Virgin Galactic's *Enterprise* sub-orbital vehicle. Rutan recently spoke at a dinner at the International Space Development Conference in Los

Angeles. He admitted that he wants to travel to the Moon. And given his track record, he will probably get there before anyone else does.

Other serious competitors from the X-Prize competition are working on inexpensive launch vehicles. A Canadian team is building the *Arrow* launch vehicle, a more traditional rocket, but with innovations that make it inexpensive, and each stage recoverable for reuse.

Arrow launch vehicle, possible future cheap access to space Courtesy CanadianArrow.com

Several other fledgling launch companies were formed to pursue the X-Prize. While I can't vouch for any of them, I think some will develop reusable sub-orbital vehicles. Some of those, hopefully, will continue on to orbital technologies. Sadly, many will probably fail, but it only takes one or two successes to revolutionize the launch vehicle industry and give us cheap access to space.

To give you an idea of how many companies are working on cheap launch vehicle technology, here's a list of names that I was able to round up (not including companies previously mentioned):

Acceleration Engineering (US)
Advent Launch Services (US)
American Astronautics (US)
Arca (Romania)
Armadillo Aerospace (US)
Bristol Spaceplanes (UK)
Discraft Corporation (US)
Flight Exploration (UK)
Fundamental Technologies (US)
High Altitude Research Corporation (US)
Il Aerospace Technologies (Israel)
Interorbital Systems (US)

Kelly Space & Technology (US)
Lone Star Space Access (US)
Micro Space Inc (US)
Pablo de Leon & Associates (Argentina)
Pan Aero Inc (US)
Rocketplane Limited (US)
Space Transportation Corporation (US)
Starchaser Industries (UK)
Suborbital Corporation (Russia)
TGV Rockets (US)
Vanguard Spacecraft (US)

As soon as relatively cheap access to space is available, then we will truly be in the age of space tourism, and our true romance with space will begin. "The allure of having sex in space will entice many people to become space tourists," Space Activist and Business Researcher Amanda Harris told me. "Private space companies must be able to meet the needs and wants of their customers to compete successfully in the space tourism industry. These companies must provide the ultimate space experience. After the excitement of seeing the curvature of the Earth wears off, as oddly as that sounds, entertainment and comfort issues are going to be of more concern to companies," she explains. "The availability of private rooms where couples can experience sex in complete weightlessness can address those issues," she says.

The business side will have to take over to make a profit. "The way to build the space tourism industry is to model it after the cruise line industry," says John Spencer. "Cruise ships are very advanced pieces of technology designed for one reason, to bring people into a unique environment, the ocean is also pretty dangerous, have a luxurious safe experience, and make a profit doing it," he says. "So it's a great model for space."

Even Spencer would agree that space cruise ships seem a long way off from becoming a reality. Firstly, in the near future, the costs will still be too high for demand to follow. Second, the staff isn't going to be available, yet.

Dr. Harvey Wichman, a leader in Aerospace Medicine, explained to me some of the realities of life in orbit. "For at least a week you can tolerate virtually anything, just for the novelty of it. It is really neat and fun. But week after week after week? You can put them through almost anything you want [for a week] and they will come back and rave about how fun it was. But the people who work there for six months and have to strap themselves to the toilet every day, oh my," he says.

Remember, space is exhausting and difficult after long periods of time there. Any hotel staff would probably be there for a longer trip, and that's when problems would begin. Rotating the hotel staff would probably greatly cut into profits. But based on psychological as well as physical reasons, the staff members may just have to be rotated every two to three months, instead of the six month staff tours the business side would prefer.

Robotic cleanup and other activities should be maximized to keep the staff to a minimum. Probably in the near term, the hotel experience will be more like a camping trip, where the tourists have to clean up after themselves and "pack out" their own garbage. Each trip would have one or two regular crewmembers acting as guide at the hotel, providing assistance, fixing emergencies, and piloting the ship up and back. But for the most part, early space hotel tourists will have to take care of themselves, because staff members will be too expensive.

In the first 89 shuttle missions there were 1777 separate medical incidents, including 141 injuries and the 7 fatalities from the *Challenger* accident. Many of these incidents were probably the results of space sickness, but many more were minor injuries, some even needing a few stitches. It would seem reasonable that space tourists will have a similar number of medical incidents, probably more unless the tourists were required to be in the same great physical shape as astronauts (which is unlikely). Therefore, a doctor or other member of the crew must have at least some medical training to handle any first aid and emergencies that occur.

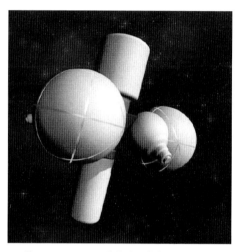

Exterior view of Orbital Super Yacht "Destiny" designed by John Spencer

Even before future cruise ships sail on the cosmic seas, John Spencer believes that the technology necessary for space tourism will be advanced by billionaires sponsoring their own private space yachts. "The yachts are a great model as a precursor to the cruise ships because they exist for experiential stuff; ego, reward, marketing, and they don't make a cash profit," he explains.

Dr. Harvey Wichman points out that insurance companies will have to get involved, and they can act as de-facto regulators. By making certain activities uninsurable, you effectively prevent them better than any government regulation could. "My bet is when insurance companies get caught up in this, one thing that is going to be demanded is that people wear dosimeters," says Wichman. "Hopefully as technology progresses and space stations can become just a little bit heavier, we'll get a little bit better radiation protection."

But John Spencer is undaunted, "A lot of stuff we talk about now is inevitably going to happen. Orbital yachting, orbital sports, cruise ships [in space], lunar resorts, we will do, because we want to. There will be disasters, but disasters don't stop progress," he explains. "Sometimes they inspire other progress."

Whatever happens with the future of tourists in space, sex will be *the* part of the equation that sells the package. "It has been said that 'sex sells,' and it does!" says Amanda Harris. "A large percentage of television commercials allude to sex even though the product being sold has nothing to do with sex. Sex is an attention-getter and a marketing tool. Sex is not a fad. It does not go out of style. Whether on Earth or in space, as long as we exist, our desire for sex will exist," she explains. "Sex will always be one of our greatest motivators."

Space Hotels, Come Fly With Me!

So while some are working on access to space, others are already working on ideas for hotels in space. One such company is the Space Island Group, which is working to fund and open a private space hotel within the next decade. The group wants to take External Tanks from the Space Shuttle, clean them out, link them up, and use the resulting structure as a space station. And they have lots of ideas for how this station will be used. On the Howard Stern show, Gene Myers, Space Island CEO, predictably found there were lots of people interested in how to have sex in space. "At first the plan was just to open a space hotel," he told Stern. "But our research shows that the real reason couples want to spend a week in space is for fabulous sex."

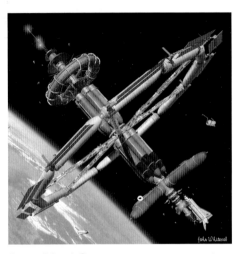

Space Island Group concept space station.
Courtesy Space Island Group

Meyers also has an interest in creating zero-g Olympics, but is still not sure exactly what the events will be. In the future, tourists may very well come to space hotels for great sports and great space sex.

Clinton Wallington, a professor at the Rochester Institute of Technology teaches a course on space tourism. "We're talking about a commercial venture here, and sex sells," he told a reporter. "If you can sell a honeymoon vacation in the Poconos, why not one in space?" Based on his advice, I would name the space hotel, 'The Zero-g Spot.'

The Honolulu Architecture firm, Wimberly, Allison, Tong, & Goo (WATG), has designed a 100 room hotel in space. This orbital station spins for partial gravity in the guest rooms, dining room, and bathrooms. Guests would probably appreciate this feature since having some gravity makes it easier to eat and use the bathroom. The core would be weightless for sports in three dimensions. In a New York Times article, Vice President Howard Wolff promised that the hotel would have private zero gravity honeymoon suites for, "really, really creative sex."

John Spencer, who has designed living spaces that have flown in space, underwater, and in Antarctica, envisions a different design. The entire hotel is in zero gravity, "Weightlessness is the whole point of the journey," he says, "so why would you want to compromise it by making it feel anything like Earth? That spoils all the fun." His hotel would be made of interconnected spheres that would be inflated in orbit.

Robert Bigelow, owner of Budget Suites of America and Bigelow Aerospace, has what seems to be the most realistic idea, at least in the short term. This is due partly to the fact that his firm is the furthest along in development. Bigelow Aerospace has several buildings at a facility on 50 acres of land at the outskirts of Las Vegas and a location near Houston Texas. The BA330 space habitat will be a 45-foot long, and 22-foot in diameter expandable module. By adding more modules, the structure would grow into a space hotel. The company has already built a mockup of the space station at the Las Vegas facility. Bigelow committed $500 million of his own money to get the habitat up and running in orbit within the next decade.

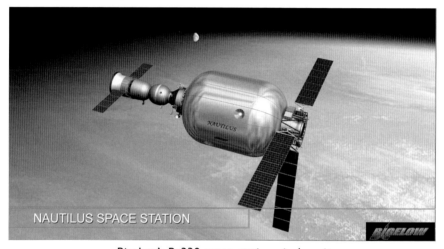

Bigelow's Ba330 space station, single unit

Interestingly, the concept of expandable technology was actually developed by NASA as part of a project called TransHab. The architecture was created by NASA Senior Engineer William Schneider who started the work in 1997. TransHab was considered for use as part of the International Space Station. However, for budget reasons TransHab was canceled in the year 2000.

According to Popular Science Magazine, Bigelow sees that as part of NASA's inefficiency, "I've put together many, many projects involving a lot of money and a lot of people," he said. And unlike NASA, "I'm used to doing things pretty darn well on budget and pretty darn well on time."

When NASA cancelled the TransHab program, Bigelow bought the exclusive development rights and signed a Space Act Agreement with NASA allowing him to collaborate of members of the TransHab team. Schneider himself had since retired from NASA, but was soon enticed to consult for Bigelow. "When I walked in here, boom! It was mind-boggling, because this is the vision that I really wanted," Schneider told Popular Science. "Here's these things, all sitting there, and of course some of them are mock-ups, but the rest were inflatable, and I said, 'Man, he's serious, He's not playing around.'"

Each module is designed with safety in mind. The layered external walls are a foot thick, include micrometeoroid shields, pressure restraint layers, and bladders to hold the air. The modules are launched in a compressed state and are very light for their size, so it doesn't take a particularly powerful launch vehicle to put them in orbit. As a result, Bigelow plans to launch his modules on inexpensive future vehicles such as the *SpaceX Falcon 9* launch vehicle or the existing Russian launchers such as the *Proton* launch vehicle.

To perform system tests, late in 2006 Bigelow plans to launch two one-third-scale modules via converted Soviet-era missiles. The modules will reach low Earth orbit and be used to test inflation, systems, radiation, space debris penetration. After several years, the modules will burn up in the atmosphere.

The outermost layer of the BA330 modules, the micrometeoroid and orbital debris (MMOD) shield, consists of five layers of graphite fiber separated by foam spacers, which has more bullet stopping power than three inches of aluminum. In tests, it has been shown to stop impacts by a five-eighths-of-an-inch in diameter aluminum pellet fired at 6.4 kilometers per second, comparable to an orbital collision. This is significantly better than any existing manned vehicle and is part of the reason that the modules are expected to last 15 years. However, getting the MMOD to fold properly for launch has been a major design headache.

Bigelow is said to be involved with every aspect of his aerospace operation. He's on the shop floor talking with machinists, and personally signing off on all of his engineers' designs. Schneider compares Bigelow to another wildly successful mogul fascinated by aerospace, "Bob is like Howard Hughes reincarnated," he told Popular Science. "He's not just a financial person; he's in the middle of everything we do."

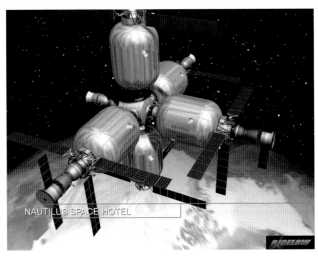

Bigelow's BA330 space hotel, multiple units

"I have little doubt that the basic [BA330] technology is likely to work," John Logsdon of George Washington University's Space Policy Institute told Popular Science. "The issue is whether there's a transport system that can get people or things, or both, up there."

Due to the concern with transportation to space, and following the lead of the Ansari X-Prize competition that was so successful in helping to develop sub-orbital vehicles, Bigelow has launched his own competition; America's Space Prize. The first privately funded vehicle that can put five people in orbit and dock with his space station, will win $50 million! The deadline for developing the launch vehicle is 10 January 2010. Bigelow reportedly wants his station to be operational for customers by 2015.

If it works, Bigelow's next giant leap will be to offer commercial services such as weeklong trips for tourists. Eric Anderson, President of Space Adventures, once told reporters that he estimated if the price is low enough you could get 20 to 30 customers per year. But Bigelow's station will be open to more than just tourists. There is probably a market for all kinds of motion picture and television producers, microgravity researchers, and even manufacturers who want to produce things that can only be made in weightlessness.

Will it work? It's probably the best shot we have for a private space station or hotel within 15 years or so. "It's a gamble," Bigelow told reporters. "But you know, the faint of heart never won a fair maiden, never won wars…I think what we're doing has some national value, win or lose."

The Heavenly Bedroom

Imagine yourself floating inside a luxurious space hotel suite decorated in warm, comforting colors and soft padded walls. Beautiful classical guitar music plays in the background. There is no "up" or "down" so all the walls have useful creature comforts available, such as a mini-bar. There is no gravity to pull you down to the floor, so you just float there in the middle of the room. You look out the huge window at the stunning Earth below. Swirling clouds, oceans, and islands seem to drift by your window. 'I've never seen something so beautiful,' you think to yourself. It has a profound effect on you. You are hyper-aware of yourself, your surroundings, and your significant other. You reach out to hold his or her hand and you pull towards one another. You lock into a close embrace, cuddling and looking out the window at the Earth and stars. The experience is pure magic, and incredibly romantic.

There are several opinions about what this ideal space bedroom should be like. Science Fiction Author and visionary Isaac Asimov described his view in a 1973 *Sexology Magazine* article, "Sex in a Spaceship." "The large spaceship of the future may contain special rooms devoted to sex," Asimov wrote. "They would be rather large to allow plenty of maneuvering. They would probably be spherical in shape, since there is no point in having distinct walls, ceiling, or floors; they would have no meaning at zero-gravity. There would be no need of furniture of any kind; in fact if present it would probably pose dangers. The spherical surface would be padded

A zero gravity sports sphere as part of a future space resort.
Artwork courtesy Shimizu Corp

with some washable material, and this might become part of the play itself. The room might be one huge trampoline." Sounds like a fun place to visit. Doesn't it?

Rod Roddenberry shared with me his idea, also a spherical design, but instead of padded walls, he would like to see the Universe outside. "You need to construct some sort of vessel that's got a glass dome so you're underneath the stars with the Earth and the Moon," he explained. "I mean you don't want to be in the circuitry and all that. You would want to be in the vastness of space, those two living life forms that become one, in the vastness of space, in a glass bubble. That would be really cool. That would be the most romantic," he admitted.

Samuel Coniglio, Photographer and Vice President of the Space Tourism Society, envisions a much smaller room than Asimov's, but agrees with Roddenberry about the view. "Based on some informal interviews with astronauts and space experts, people tend to congregate toward smaller, more intimate spaces," he explains. "On the *ISS* for example, the crew spends most of their time in the smaller Russian module as opposed to the more spacious American modules. People want to go to space for two reasons: zero gravity and the fantastic view. When you are on the *ISS*, the first is automatic, but the second is limited to tiny view ports," he explains. "People will want a panoramic view not unlike one sees on the observation deck on the Starship *Enterprise* in *Star Trek*: the Next Generation."

View of the Earth from Space. The large circular shadow on the
Mediterranean is the shadow of the moon during a total Eclipse

But window sizes have to be realistic. At room pressure, 15 pounds per square inch, the atmosphere pushes with over a ton of force on each square foot of window space (15 x 144 = 2160 lbs). So a large picture window made out of glass isn't possible. But other clear materials like thick sheets of Lexan are possible. And a dome (or hemisphere) shape is stronger structurally than a flat panel. There may be support ribs, or even exterior steel cable netting to reduce the load on the window. This would break up the view into smaller areas, like the seams on a soccer ball, but the overall effect would still be a grand window. As long as there is a good sized overall view, hotel guests will be happy, and can live with these drawbacks. The windows would of course be polarized, and would ideally have a layer which could dim or black out when desired by running a current through. Otherwise, the happy couple would be awakened every 90 minutes with another, albeit beautiful, sunrise.

My Ideal Space Bedroom

My ideal space hotel suite would consist of a few different rooms, a lounge, a bathroom, and most importantly a bedroom with an incredible view of the Earth and stars. Visualize a sphere, maybe eight feet in diameter. The view dome is on one hemisphere and the padded "bed" wall is on the other. A single door hatch towards the side opens to the rest of your private suite. The surface of the padded wall has short puffy threads for Velcro-like shoes and other objects to stick onto. All along the padded wall are straps and tethers to tie things onto, like sleeping bags. Vents between the padded wall and dome window would circulate the air. You could use a sleeping bag for two (see Chapter 2, *How to Make Love in Space*), or just float in the air, bumping against the sides as you gaze out at the heavenly view.

Since free liquids are a problem in space, and passengers cannot be counted on to keep all of their fluids (Martini's or whatever) contained, guest bedrooms are free of electronics. Audio sensors in the air vents listen for your voice commands to dim the window, or turn on the music that you've had sent up electronically from Earth. Your CD collection weighs too much to carry, so your top one hundred albums were sent by electronic signal to the hotel computer. Remember that signals are cheap to send, items with mass are not. There are filters over the air vents to catch any stray liquids, from the wine squirt bottles, or even other, more intimate fluids, and draw them away via surface tension. The fans are placed quite a ways down the vent tubes so that the room is as quiet, peaceful, and heavenly as possible.

Just outside the hatch is a robotic cleaner waiting patiently to enter after you leave for dinner so that it can scrub the walls and clean up after you and your companion. Based on the robotic vacuum cleaners that have been around for years, these robots are much more sophisticated. They can grab onto the walls and hold on as they clean. Remember that "up" and "down" are meaningless in space.

Beyond the bedroom access hatch, in a separate area, is the dreaded space toilet. Guests hate using it, because it's really a vacuum cleaner for your waste with seat straps. It's simply disturbing. What misses the hole has to be wiped up by hand. Next to it is a large supply of antibacterial wipes from hand size to full body size. You use them for all cleaning because there are no sinks or showers possible in space. The hotel staff pre-briefed you on how everything works, but even so, every area of the suite, including the bathroom, has printed directions on the wall in case you forget or get flustered. Looking at it, you're reminded of the scene in the movie *2001*, where Dr. Floyd spends time reading the directions for the zero-g toilet!

With several of these bedroom pods facing outwards, no guest would be able to see any other pod, thus ensuring privacy. Space Architect John Spencer explains his goals when designing a space bedroom, "We want to create a luxurious environment, privacy; the sensuality of having noise control is an important issue. Our job as designers is to create these beautiful sensual safe environments so it's enhanced into the whole lovemaking process."

From the outside, our hotel might look like a collection of silver soap bubbles, each bubble a private bedroom with a mirrored surface for privacy. Some rooms might have special themes, like a hydroponics garden, or an interior room with images shown on all of the walls so that you can experience an exotic view like the rings of Saturn, or colorful nebulae all around you.

Some couples may want to record their intimate rendezvous activities for later home viewing to remind them of the experience. Some might even want to show their video to others or sell it; there is certainly a market for such voyeurism. Even if dozens of these videos are on the market, based on the success of the adult entertainment industry there always seems to be a demand for more. So at opposite ends of the room, behind sealed windows, small video recorders could be turned on or off as the couple requested. The cameras are hooked to a small computer that isn't networked to the rest of the station to ensure that only the occupants have access to the original digital recording.

And there are some great spin-offs from developing this romantic bedroom. John Spencer says, "If we can create sensual environments in space, then we've solved so many problems such as space sickness, safety, confinement, noise, privacy. So many things need to be addressed so that people are comfortable enough to want to do this, and secure enough to do it, and do it a lot. That means we've solved so many issues that we've really created a leap forward in creating a better space experience," he explains.

"When it's a very comfortable environment to have sex in space," Spencer continues, "lovemaking in zero gravity, we've solved a whole lot of other issues that are enhancive to that. And if it's enhancive to that, then it's enhancive to creativity, to other issues, to even scientific research. If you are more comfortable and open-minded because you are in a creative, safe environment, it will stimulate inventions and ideas and perceptions. All of those things are what the real value is for space. It's not mining the Moon, or Asteroids, or solar power satellites, it's really that it's a new realm for humans to explore and be inventive and create stuff," he says.

Imagine yourself and your significant other in this heavenly chamber. You float in a state of bliss, the music uplifts you and you fly into space together holding on to each other, with no feeling of motion except a mild breeze blown into the room as the Earth moves below and the stars shine above.

The Future

One day far in the future we will not only have hotels in low Earth orbit, but hotels on the Moon. Eventually we might even have orbital colonies and habitats on the Moon. But first we have to get there.

Right now it takes about six months to get to Mars. Without a major technology breakthrough, that probably won't change. So the round trip mission would last at least two years, and probably longer. Will bored astronauts play on their long trip to and from the Red Planet? I would think so.

Some people think that the current vision of a single ship flying to Mars is too limited. If robots do all of the early exploration, then the first human voyage to Mars could be much more ambitious. "It won't be one little ship with six guys in a can," says John Spencer. "It will probably be a dozen ships. No joke, I predict a hundred people on the first human mission to Mars. We're not going to do stuff the way people think we are going to do it," he explains. "It's all going to be sponsored by international competitions, media, and stuff like that."

Settlement of the Moon may happen much faster than settlement anywhere else in our Solar System, possibly even before the first humans venture to Mars. The initial exploration missions planned by NASA could easily grow into larger commercial ventures. "A Moon hotel would be much, much easier [than an orbital one]," says Dr. Harvey Wichman. "What we need to find out is; is the one sixth gravity on the Moon sufficient to ameliorate the effects of reduced gravity? We don't know that," he explains. "And that's why we need to get up to the Moon and start doing some research."

No matter where on the Moon that you place a lunar base, at least part, if not all of it, should be underground to protect inhabitants against high radiation while sleeping, and during a solar storm. If the base was underground, then to keep it from being dark and depressing, there would have to be a way to let light in. Perhaps giant pieces of glass or silicon crystal could be made on the Moon and used in the construction on the habitat. The mega crystal(s) could be placed so that even deep underground, inhabitants could experience reflected natural light, at least during the daytime. If the base was built on a polar peak, the light would always be streaming in as there are peaks and ridgelines where the Sun never sets.

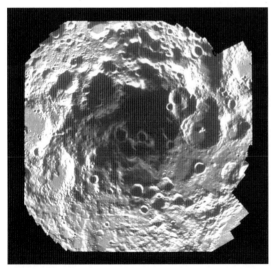

These spots are also ideal to continually generate solar power, just keep rotating the array once every 28 days to track the Sun. A dozen kilometers or so below the polar peaks of eternal sunshine are the craters of eternal darkness where water may lie.

Lava caves created by the Moon's formation may be commonplace and strong enough to use as a human habitat. In a New York Times article, Boeing Engineer Gregory Bennett proposed, "Just seal off and pump in air for an instant Lunar habitat... The cave would

Clementine mosaic of the Lunar south pole. The black is the Crater of Eternal Darkness, the bright area above it, the Ridge of Eternal Sunshine have the additional benefit of providing protection from nasty radiation," he explained.

"Down the road when we have facilities on the Moon," says Space Psychologist Sheryl Bishop. "It will be an interesting place to go to because you have one-sixth G, and you have other things that you can experience that will be different from your normal life back home. You can walk, but walking is different than back home. You can swim, you can do all the things that you do back on Earth, but in a significantly different gravity environment," she explains. "So it's all different."

One possible Lunar habitat, notice that it's covered by regolith

In the future, a broader range of humans will be attracted to live and work in space. More work will be done there, including the design, construction, and maintenance of new habitats. So naturally there will be more opportunities for sex, and space workers will be observed less than the small number of astronauts that currently go on missions into space. This will lead to families in space, and definite risks, and possible changes, to the child born in a non-Earth habitat.

Once people have lived on the Moon together for a while, attachments are bound to form and families will be created. That assumes that one-sixth gravity is sufficient for conception, and childhood development. But children born and raised on the Moon may be prisoners of space. They may not be able to adapt to Earth because their bones and muscles may not be strong enough to tolerate Earth gravity.

Children of the Moon or Mars may be exiles of Earth. Unless we find a way to strengthen muscles, bones, and balance systems, these poor children would never be able to walk on the surface of the Earth. Personally, I couldn't do that. I'd want my child to have every opportunity to return to Earth and go running, jumping, swimming, and bicycling if he or she wanted too. The Universe should be open to all children, to all people.

Some believe that future medical technology will be able to overcome and reverse the limitations of a body that develops in low gravity. Space Advocate Derek Shannon says, "I think that environments of varying gravities will only be a small part of the broader changes in living and reproduction as humanity masters genetics and nanotechnology," he explains. "But these technologies will also mean that successful adaptation to new environments will not only be possible, it will be reversible as necessary, so that a baby born in zero-g could still climb Olympus Mons or Mount Everest if she wanted. Assuming the kid wouldn't prefer the virtual version of such an experience, of course."

Author, Musician, and Space Activist Vanna Bonta says, "It's vital as we postulate and work toward exploration and human settlement beyond Earth. I like to think of the possibilities of sustaining humanity's continuum, with preserved recorded history way beyond the life of our Sun."

Future orbital colony as envisioned by Space Artist Don Davis (for NASA)

Our Solar System is filled with resources that we can use to create wonderful human habitats. The asteroids and moons of the Solar System will someday provide the resources for building in space. Space mining will eliminate the need to haul up expensive fuel, concrete, water, air, glass, iron, copper and other metals and products necessary to build habitats and sustain life. It's all there waiting for us. We need to develop the capability to mine it and change the raw materials into something useful.

"My hope, like any good Trekkie, is that one day we will be spread out over several planets," says Aerospace & Human Factors Engineer Juniper Jairala. "I don't really know if people intend to live in space. Humans as a species are interested in going to other planets and meeting other species. I certainly want to hang out in space," she admits. "But I don't know if I want to be out there for a long time... If we don't find a way to colonize Mars and the Moon and make it worthwhile, and get enough interest to live on those planets, because they are kind of desolate and far. But we have to get off of Earth. We might end up on large orbiting space colonies. I think that we should have both," she says.

The future of human expansion into space looks promising once the technology makes it feasible. Samuel Coniglio put it this way, "Once we get over the religious, political, technical, and psychological hurdles, our society will see this expansion to the stars as inevitable," he says. "Space Travel has little to do with technology as it does expanding the human spirit."

Settling in Space

I think Samuel Coniglio got it right when he said the human spirit is what drives our desire to explore and expand into space. Vanna Bonta put it this way, "Human migration beyond Earth is our destiny, not as in something preordained, but in the sense of it being inevitable and necessary. The lot of humanity seems to be that of expanding consciousness into new territory."

Sex is needed to settle the galaxy. We are each the product of thousands of generations of humans having sex. In the long run, the Earth may be too dangerous a place for people to survive. Several great extinctions have happened on Earth, and they will happen again. Perhaps as an intelligent species we can prevent them. Or we can grow beyond them by living in habitats and on planets beyond Earth. Or, we can do both. There are many possible threats out there, some we impose on ourselves, like nuclear winter, pollution, or population overcrowding, and some are external like asteroids. But space, whether through the development of new technology or settlement of new habitats, can give us possible ways to escape almost all of the bad stuff.

We need to be smart about this expansion into space and not trample others in our exuberance. "I think it's wise to heed the caution of [Science Fiction Author] Kim Stanley Robinson," says Derek Shannon. "He reminds us how hurtful 'Manifest Destiny' and similar jingoism have been throughout history. We're not destined to expand humanity or intelligence into the Universe, but it is an option available to us. Rather," says Shannon, "it is imperative that we create the best possible future for all humankind, and settling the Moon, Mars, and beyond can definitely be an important part of that."

In the future, humans will change. Evolution and natural selection takes a long time, but will still affect us. However, with new biomedical developments, we are gaining the ability to help ourselves and change our own genetic makeup. Already doctors can screen for diseases even before a child is born. Someday we will be able to change ourselves to adapt to other environments so we can live almost anywhere. This will greatly accelerate the process of natural selection and allow us to live in extreme environments that we find impossible to live in now, or ones that we cannot even imagine today. I believe that humans will expand on a broader scale, and differently than anyone can foresee now. In the future, humans will adapt to their alien environments. It's amazing when we stop to consider that we *are* the parents of space aliens.

Sexy Science Fiction

Space is sexy. It pushes our primal buttons. It's a power trip. That's why it's so much a part of Science Fiction. Science Fiction is always weird, and almost always, I'd say 90% of the time, has something to do with sex. Science Fiction writers have always tried to predict the future. Sometimes they're right on target, and other times they're way off. In any case, we love Science Fiction because it gives us an exciting glimpse of what the future may be like for us, and for our children, as science and technology move forward. There will always be some things about future society that we can't even begin to predict. That's what makes the future fun, scary, and exhilarating too. The excitement of space and the future always seem to lead back to sex.

Sex in Science Fiction is too broad of a topic to cover in one short section of this book. The various views of sex and society in Science Fiction are as varied as the possible realities portrayed. But, I wouldn't feel that this book is complete without at least briefly discussing sex in science fiction.

First off, I'm confident that Science Fiction protects you from "future shock." If we didn't have Science Fiction to prepare us for strange futures, our ever-changing technological and social scene would leave us spinning. Second, Science Fiction reflects the issues of the time and makes us look at ourselves now without discussing the issue directly. So sex in Science Fiction is used not only just to sell the story, but also to look at the sexual attitudes of our own society. And to prepare us for possible significant changes in sexual attitudes that may come in the future – possibly a future much closer than the story depicts.

Sex in Science Fiction goes way back, but let's start in 1900. That was the year that George Griffith wrote what I think is the first book about sex in space. *A Honeymoon in Space* follows a newlywed husband and wife on a grand tour of our Solar System. Many believe the entire work is a metaphor for sexual intercourse.

Edgar Rice Burroughs' series on "Barsoom," or Mars, was first published in 1917, and began with the sexually suggestive, *The Princess of Mars*. The romance between John Carter and princess Dejah Thoris was slightly racy for the time, perfect for its target audience of titillated teenage boys.

The Fantasy book *Jurgen,* published in 1919 by James Branch Cabel, had such a strong sexual theme that it was banned in many places and the author was prosecuted for obscenity.

Aldous Huxley's groundbreaking book, *Brave New World,* published in 1932, showed us a world where sex is only used for promiscuity, not procreation. George Orwell's book *1984* was published in 1948 and showed a state that severely and ruthlessly restricted sex. Part of Orwell's warning was against legislating sexual behavior, such as fellatio and anal sex, something that some states in the United States did on "moral" grounds.

The Science Fiction pulp comics of the 20th Century exploited sex, at least on the covers. There are frequent pictures of scantily clad women. The cliché is a nearly naked woman being captured by aliens, or space pirates, or other unsavory characters. Although why an alien would want a human woman, or man, is beyond me. We would be as ugly to them, as they would be to us. As you look at more and more comic and magazine covers, you notice that a significant number show women in strong roles, clearly in charge of the situation, holding ray guns or the like, even if scantily clad. So perhaps they were not such bad examples after all. Besides, if women were in charge, they surely decided, maybe even designed those revealing costumes.

Science Fiction movies have always included sex in the story, because sex sells. Some, like the 1960s campy classic *Barbarella,* have overt nearly nude shots, outrageously sexy costumes, and discuss sex openly. In the original movie trailer for the futuristic *Barbarella,* the title character, played by Jane Fonda, is a "Five star, double rated, Astronavigatrix Earth girl who's specialty is love." She is smart enough to pilot her own starship and save the Universe, but very naïve when it comes to sex. Apparently no one on Earth has had sex in centuries, only barbarians do *that*. Shortly after crash landing on the target planet, she becomes something of a nymphomaniac.

Other more mainstream American movies show romantic banter and sexual tension without exposing anything. Take for example the sexual tension between *Star Wars* heroes Han Solo and Princess Leia, which never gets hotter than a big kiss on screen.

In the recent science fiction movie, *Supernova*, especially on the uncensored DVD release, you can see what the director envisioned as sex in zero-g. There are a couple of sexual space encounter scenes in which characters float in an observation dome oddly sticking out from the middle of the ship. Although filmed here on Earth, and heavily edited with computer animation, these scenes give a pretty good idea of what a romantic zero-g encounter might be like. Besides, the romantic space music is great.

The 1956 Science Fiction classic movie, *Forbidden Planet*, had a sexually repressed, all male crew that was hot for the female lead, Altaira Morbius. Altaira is the naïve daughter of the mysterious Dr. Morbius, smart, attractive, with a book knowledge of sex, but physically and emotionally unaware of her own sexuality.

In *Alien*, Sigourney Weaver showed us Lieutenant Ripley, a strong female character who was a tough survivor, while still being great looking. It definitely shattered the 'women are weak' stereotype.

For outrageous future fashion, nothing beats the 1997 action movie with a sexually suggestive theme, *The Fifth Element*. In it Leeloo is the perfect woman who must save the Universe from evil. In the end she is only able to do it because of love.

Science Fiction on television usually consists of a strong lead character that inspires sexual fantasies. The *Star Trek* captains and many of the supporting characters, both male and female, are obviously chosen for sex appeal. This includes Captain Kirk, Picard, Spock, Riker, Troi, etc. The late 1970s show, *Buck Rogers in the 25th Century*, revolves around the characters of Buck Rogers and Willma Deering, both young, fit, sexually attractive astronauts. The show basically played the 'will they or won't they?' game with viewers.

Robert Heinlein opened Science Fiction to group sexuality and marriage in the 1960's with his books, *Stranger in a Strange Land* and *The Moon is a Harsh Mistress*. Also, I'm told that Theodore Sturgeon's 1960's book *Venus Plus X* was groundbreaking with its theme of transsexuality, but I have not been able to read a copy myself.

When researching this book, several people brought up Kim Stanley Robinson's book *Red Mars* and its sequels, *Blue Mars* and *Green Mars*. My husband has read it and it's on my long list of books to read. Basically, the book deals with the first settlers on Mars, and is filled with realistic science, and human behavior including personal conflict, politics, psychology, culture, and of course sex plays a significant role.

Harlan Ellison edited a Science Fiction anthology of taboo topics including homosexuality, prostitution, drugs, being sexually neutered, and a host of other topics, with a series of short stories in *Dangerous Visions* in 1967, and the 1972 sequel *Again, Dangerous Visions*. The third anthology, *The Last, Dangerous Visions* was supposed to be published in 1973 but for various reasons the book has never been published.

Several books look at the alien sex, such as the 1949 book *Venus and the Seven Sexes* by William Tenn. In it, Tenn describes seven different sexes needed on Venus for procreation. Theodore Sturgeon's 1953 book, *The World Well Lost*, discusses alien homosexuality; probably to cast a mirror on society's own view of homosexuality at the time.

Sex between humans and aliens is a strange concept to me. I can't see what the attraction would be, but then to each his/her/its own. In *Ringworld Engineers*, Larry Niven even comes up with a term for it. "Rishathra" is used to mean sex between individuals of two different humanoid species. One advantage of rishathra is to provide sexual release without the risk of pregnancy. Science fiction fans have spread the term. Apparently, there is no other single word to describe sex with a space alien.

Although *Star Trek* used the idea of sex between different humanoid species, writers took the liberty of allowing cross species pregnancy. This is obviously unrealistic for nature to perform by itself. However, in the future, it might be possible via genetic engineering to combine the DNA from two species to form a new species that's a conglomeration of the two different humanoids.

I like the *Star Trek* idea of infinite diversity in infinite combinations, and I think that it applies to sex. To quote from a *Star Trek: The Next Generation* episode titled *The Outcast*, "What makes you think that you can dictate how people love each other?" No matter how we humans may change when we settle the galaxy, love will always be Universal.

Conclusion: Sex to Settle the Galaxy!

"The Earth is just too small and fragile a basket for the human race to keep all its eggs in." - Robert Heinlein

I'm what Space Architect John Spencer describes as a Primal Futurist. This is a person who uses high technology to fulfill the basic primal urges and dreams that are hardwired into us; exploration, procreation, and expansion of the species. Space is where humanity will go, and our primal urges for survival, sex, and curiosity, will keep us exploring the unknown. Love makes the Universe work, and gives us the power to create the future. Love and space feed our souls.

I have found that sex and reproduction are almost never mentioned when talking about our future in space. Since we're at the beginning of the space tourism era, I'm sure that someday soon people will have sex in space, and there are some serious issues to explore; sexual health, pregnancy, and childbirth. In the preface to his doctoral thesis on human sexuality and spaceflight, Ray Noonan put it best, "Wherever humanity goes, our sexuality will surely follow." So we'd better be ready to deal with the consequences of the nature of humanity. The sex drive rules all!

Vanna Bonta expressed it well, "People have been making love and having sex in space over the thousands of years that our ancestors lived and traveled in small hunting-and-gathering bands. Earth is in Space. Sex in space is not about going somewhere else to have sex; it's ultimately about expanding beyond our immediate neighborhood, into a Universe to which we belong."

Sex in space is going to be unusual and interesting. I dream of being able to float with my husband in weightlessness taking in the spectacular views and becoming intimate in this new environment. Ultimately the novelty might wear off, however, and the difficulties of zero-g might start to become an issue. In the long run, Earth may prove to be easier, and perhaps even the most "sensual" planet. As Rod Roddenberry said, "Gravity is good, and the less gravity, the more difficult. So for a good time go to Earth."

I hope *Sex in Space* has aroused your interest, and satisfied your curiosity, about the benefits and consequences of lovemaking on the final frontier. I took pleasure in writing this book for you and I hope you too feel satisfied by *Sex in Space*.

Appendix

Note: This is NOT an actual NASA document, but a gag report found on the Internet at several sites, and was circulated extensively by e-mail. I have no idea who the author is, and therefore I can't credit he/she/them. Even though this isn't a real NASA publication, it's done in the same dry, official style as typical NASA documents. It's interesting that the author chose to portray space sex experiments in this way, since sex is obviously not a dry subject.

Experiment 8 Postflight Summary: NASA publication 14-307-1792

Abstract

The purpose of this experiment was to prepare for the expected participation in long-term space based research by husband-wife teams once the US space station is in place. To this end, the investigators explored a number of possible approaches to continued marital relations in the zero-G orbital environment provided by the XXXXXX shuttle mission.

Our primary conclusion is that satisfactory marital relations are within the realm of possibility in zero-G, but that many couples would have difficulty getting used to the approaches we found to be most satisfactory.

Introduction

The number of married couples currently involved in proposals for long- term projects on the US space station has grown considerably in recent years. This raises the serious question of how such couples will be able to carry out normal marital relations without the aid of gravity.

Preliminary studies in the short-term weightless environment provided by aircraft flying on ballistic trajectories were sufficient to demonstrate that there were problems, but the duration of the zero-G environment on such flights is too short to reach any satisfactory conclusions. Similar experiments undertaken in a neutral buoyancy tank were equally inconclusive because of the awkwardness of the breathing equipment.

The primary conclusion that could be drawn from these early experiments was that the conventional approach to marital relationships (sometimes described as the missionary approach) is highly dependent on gravity to keep the partners together. This observation lead us to propose the set of tests known as STS- 75 Experiment 8.

Methodology

The co-investigators had exclusive use of the lower deck of the shuttle XXXXXXXX for 10 intervals of 1 hour each during the orbital portion of the flight. A resting period of a minimum of 4 hours was included in the schedule between intervals. During each interval, the investigators erected a pneumatic sound deadening barrier between the lower deck and the flight deck (see NASA publication 12-571-3570) and carried out one run of the experiment.

Each experimental run was planned in advance to test one approach to the problem. We made extensive use of a number of published sources in our efforts to find satisfactory solutions see Appendix I), arriving at an initial list of 20 reasonable solutions. Of these, we used computer simulation (using the mechanical dynamics simulation package from the CADSI company) to determine the 10 most promising solutions.

Six solutions utilized mechanical restraints to simulate the effect of gravity, while the others utilized only the efforts of the experimenters to solve the problem. Mechanical and unassisted runs were alternated, and each experimental run was videotaped for later analysis. Immediately after each run, the experimenters separately recorded their observations, and then jointly reviewed the videotapes and recorded joint observations.

The sensitive nature of the videotapes and first-hand observations precludes a public release of the raw data. The investigators have prepared this paper to summarize their results, and they intend to release a training videotape for internal NASA use, constructed from selected segments of the videotapes and additional narrative material.

The following summary is organized in two sections; the first covers the mechanical solutions, while the second covers the "natural" approaches. Each solution is described briefly, and then followed by a brief summary of the result. Some summaries are combined.

Summary of Results

1) An elastic belt around the waist of the two partners. The partners faced each other in the standard or missionary posture.

Entry was difficult and once it was achieved, it was difficult to maintain. With the belt worn around the hips, entry was easy, but it was difficult to obtain the necessary thrusting motion; as a result, this approach was not satisfactory.

2) Elastic belts around the thighs of the two partners. The female's buttocks were against the groin of the male, with her back against his chest.

An interesting experiment, but ultimately unsatisfactory because of the difficulty of obtaining the necessary thrusting motion.

3) An elastic belt binding the thighs of the female to the waist of the male. The female's buttocks were against the male's groin, while her knees straddled his chest.

Of the approaches tried with an elastic belt, this was by far the most satisfactory. Entry was difficult, but after the female discovered how to lock her toes over the male's thighs, it was found that she could obtain the necessary thrusting motions. The male found that his role was unusually passive but pleasant.

One problem both partners noticed with all three elastic belt solutions was that they reminded the partners of practices sometimes associated with bondage, a subject that neither found particularly appealing. For couples who enjoy such associations, however, and especially for those who routinely enjoy female superior relations, this solution should be recommended.

4) An inflatable tunnel enclosing and pressing the partners together. The partners faced each other in the standard missionary posture. The tunnel enclosed the partners roughly from the knees to waist and pressed them together with an air pressure of approximately 0.01 standard atmospheres.

Once properly aroused, the uniform pressure obtained from the tunnel was sufficient to allow fairly normal marital relations, but getting aroused while in the tunnel was difficult, and once aroused outside the tunnel, getting in was difficult. This problem made the entire approach largely unusable.

5) The same inflatable tunnel used in run 4, but enclosing the partners legs only. The partners faced each other in the missionary position.

6) The same inflatable tunnel used in run 4, but with the partners in the posture used for run 2.

Foreplay was satisfactory with both approaches; in the second case, we found that it could be accomplished inside the tunnel, quite unlike our experience with run 4. Unfortunately, we were unable to achieve entry with either approach.

A general disadvantage of the inflatable tunnel approach was that the tunnel itself tended to get sticky with sweat and other discharges. We feel that the difficulty of keeping a tunnel clean in zero-G makes these solutions most unsatisfactory.

7) The standard missionary posture, augmented by having the female hook her legs around the male's thighs and both partners hug each other.

8) The posture used in run 3, but with the female holding herself against the male by gripping his buttocks with her heels.

Initially, these were very exciting and promising approaches, but as the runs approached their climaxes, an unexpected problem arose. One or the one or the other partner tended to let go, and the hold provided by the remaining partner was insufficient to allow continued thrusts. We think that partners with sufficient self-control might be able to use these positions, but we found them frustrating.

9) The posture used in run 2, but with the male using his hands to hold the female while the female used her heels to hold the male's thighs.

Most of the responsibility for success rested on the male here, and we were successful after a series of false starts, but we did not find the experience to be particularly rewarding.

10) Each partner gripping the other's head between their thighs and hugging the other's hips with their arms.

This was the only run involving non-procreative marital relations, and it was included largely because it provided the greatest number of distinct ways for each partner to hold the other. This 4 points redundant hold was good enough that we found this solution to be most satisfactory. In fact, it was more rewarding than analogous postures used in a gravitational field.

Recommendation

We recommend that married couples considering maintaining their marital relations during a space mission be provided with an elastic belt such as we used for run 3 (see Appendix II). In addition, we advise that a training program be developed that recommends the solutions used in runs 3 and 10 and warns against the problems encountered in runs 7 and 8.

We recognize that any attempt by NASA to recommend approaches to marital relationships will be politically risky, but we feel that, especially in cases where long missions are planned, thought be given to screening couples applying to serve on such missions for their ability to accept or adapt to the solutions used in runs 3 and 10.

Resources

More resources are available on my website (www.SpaceGoodies.com), but here are some to get you started.

Books and Magazines:

Getting Off The Planet by Dr. Randall Chambers (ISBN 1894959205)

Dr. Ray Noonan's Dissertation: *A Philosophical Inquiry into the Role of Sexology in Space Life Sciences Research and Human Factors Considerations for Extended Spaceflight*

Space Tourism: Do You Want to Go? By John Spencer (ISBN 1894959086)

Women Astronauts by Laura S. Woodmansee (ISBN 1896522874)

Women of Space: Cool Careers on the Final Frontier by Laura S. Woodmansee (ISBN 1894959035)

Web Sites:

Laura S. Woodmansee (www.SpaceGoodies.com)
CG Publishing Inc (www.cgpublishing.com)
Laboratory of Cell Growth (www.labofcellgrowth.com)
NASA (www.nasa.gov)
National Space Society (www.nss.org)
Dr. Ray Noonan's Home Page (www.bway.net/~rjnoonan/)
Personal Spaceflight (www.personalspaceflight.info)
Space Adventures (www.spaceadventures.com)
Space Frontier Foundation (www.space-frontier.org)
Space Generation Foundation (http://www.spacegeneration.org/usa)
Space Island Group (www.spaceislandgroup.com)
Space Tourism Society (www.spacetourismsociety.org)
X-Prize Foundation (www.xprizefoundation.com)
Yuri's Night (www.yurisnight.net)
Zero-G Corporation (www.nogravity.com)

Acknowledgments for *Sex in Space*

My deepest gratitude goes to everyone who interviewed, or otherwise contributed to this racy little book. Thank you for helping me to make my *Sex in Space* book a reality! Including, but not limited to: Sheryl Bishop, Vanna Bonta, Samuel Coniglio, Mary Lynne Dittmar, Marianne Dyson, Amanda Harris, Millie Hughes-Fulford, Juniper Jairala, Kenny Kemp, Karen Lau, Lorelei Lisowsky, Rosaly Lopes, Regina Lynn, Ginny Mauldin-Kinney, Noah McMahon, Margaret Michels, Ray Noonan, Janet T. Planet, Eugene "Rod" Roddenberry, Karen Rugg, Vanessa Satterfield, Derek Shannon, James Spellman Jr., John Spencer, and Harvey Wichman. I'd also like to thank the Yuri's Night Los Angeles Team, the National Space Society, the Space Frontier Foundation, and the Space Tourism Society for helping with the creation and promotion of this book and its goals. And, I'd like to thank my publishers at Apogee Books/C.G. Publishing, Richard Godwin and Robert Godwin for having the courage to publish yet another one of my books, and Rick Tumlinson of the Space Frontier Foundation for writing the Foreword and for his support. Thank you to Geoff Godwin for creating a beautiful cover for this book. My infinite thanks go to my loving friends and family who have done so much to encourage me, especially my wonderful husband Paul. I love you!

Also by Laura Woodmansee:
Women Astronauts
Women of Space: Cool Careers on the Final Frontier

Praise for *Women Astronauts*

"Young women considering the sciences, and particularly looking toward NASA and space as their future, should take time to read this book, and its successor, *Women of Space: Cool Careers on the Final Frontier ... 5 Stars.*" - Curledup.com

"... This book underscores that women have become a permanent and important part of the space program, and that girls growing up today can realistically dream of becoming astronauts themselves." - Sally Ride, America's first woman in space

" . . . Warmly welcomed and recommended and with the CD-ROM it is excellent value." - Spaceflight - British Interplanetary Society

"The book will be a valuable addition to any exploration library, and should be especially important to girls and young women . . ." - SB&F (Science Books & Films)

"This is a fascinating book . . . a very thorough piece of work, and a valuable resource ..." - The Observatory Magazine

"It received our highest rating!" – AAAS Science Books and Films

Praise for *Women of Space: Cool Careers on the Final Frontier*

"Oh. Excuse me. I was just looking about for the nearest time machine. Why? Because I would like to turn the clock back twenty years or so, and have a guidance counselor or perhaps one of my teachers hand me this book. ...Inoculate your daughter against the dreaded disease of career directionlessness - get this book ... 5 Stars." - Curledup.com

"Reading [*Women of Space: Cool Careers on the Final Frontier*] was exciting, even to someone who has already made a firm decision about her career! It reminded me of all the things I like about my job — the things that I usually forget about in the midst of data reduction." – The Observatory

"Women seeking career guidance may find *Women of Space* to be especially useful. And readers who simply enjoy learning of the accomplishments of women will find each of these books immensely rewarding." – Margaret Reilly, Association for Women in Science, *AWIS Magazine*

For further information about
Laura S. Woodmansee and her books
visit www.cgpublishing.com

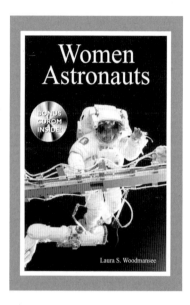

Women Astronauts
ISBN 1896522874

" . . . warmly welcomed and recommended and with the CD-ROM it is excellent value."
Spaceflight - British Interplanetary Society

"The book will be a valuable addition to any exploration library, and should be especially important to girls and young women . . ."
SB&F (Science Books & Films)

"This is a fascinating book . . ."
" . . . a very thorough piece of work, and a valuable resource . . ."
The Observatory Magazine

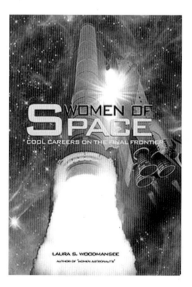

Women of Space
ISBN 1894959035

"Reading [*Women of Space: Cool Careers on the Final Frontier*] was exciting, even to someone who has already made a firm decision about her career! It reminded me of all the things I like about my job — the things that I usually forget about in the midst of data reduction."
The Observatory